COLLINS BRITISH TREE GUIDE

COLLINS BRITISH TREE GUIDE

OWEN JOHNSON

ILLUSTRATED BY
DAVID MORE

William Collins
An imprint of HarperCollins*Publishers*
1 London Bridge Street
London SE1 9GF

WilliamCollinsBooks.com

First published in 2004 as *Collins Tree Guide*
This amended edition published in 2015

21 20 19 18 17 16 15
10 9 8 7 6 5 4 3 2 1

ISBN 978-0-00-745123-4

A catalogue record for this book is available from the British Library.

Edited and designed by ketchup/Tom Cabot
Printed and bound in Hong Kong by Printing Express

CONTENTS

INTRODUCTION

Identifying a Tree

Simply thumb through the illustrations until you find a match for your tree. The text will then point you to any other trees with similar characteristics.

Larger genera are introduced by 'Things to Look For': a quick-reference list of particular diagnostic features.

Species Descriptions

Symbols Used

Key Species (for major genera): the commonest or most distinctive forms, to which all the others are compared.

The tree is (more or less) evergreen.

The tree carries its leaves and leaf-buds in opposite pairs (or threes) on extension shoots. (All trees without this icon carry them one by one.)

The tree is in part poisonous. (The toxicity of many rarer trees in this book is unknown.)

Most species descriptions include the following details:

- Preferred common name.
- Current scientific name.
- Alternative/illegitimate common and scientific names in parentheses at the start of the first line of the description.
- Basic natural distribution, including status in Britain and Ireland (native, naturalized).
- The date of the first known introduction of a species to Britain, the first occurrence of a hybrid, or the first distribution of a cultivar.
- Frequency in Britain and Ireland: Dominant (the principal tree species in an area); Abundant (locally ubiquitous and likely to be found in most places); Frequent (present in small numbers in most areas); Occasional (individuals in many large gardens or towns but absent from the majority of parks or woods); Rare (likely to be absent from many counties and found in only a few parks and gardens in others).
- Scientific family name (always ending -aceae), included in the species description if the tree is the first one to be treated in a new family and the family is not widespread

enough to warrant an introductory paragraph. Families are covered consecutively through the book, ordered by probable evolutionary relationships.

- The tree's **APPEARANCE** (the most diagnostic features are *italicized*): **SHAPE** The height given is that of the tallest specimen currently recorded in Britain/Ireland. **BARK** Of a mature trunk unless stated. **SHOOTS** Colour and appearance by the first winter (most start green). **BUDS** Colour and appearance (of leaf-buds, unless stated) during winter. **LEAVES** Note that examples can always be found whose size exceeds the suggested size parameters. (The 'bottom' of the leaf is the end nearest the stalk.) **FLOWERS** and **FRUIT** (or **CONES**).
- **COMPARE** Cross-refers to trees elsewhere in the book with which confusion is possible.
- **VARIANTS** Subspecies, varieties and cultivars that are widely grown or readily separable.
- **OTHER TREES** Some rarer allies (often not illustrated).

The illustrations show leaves, flowers and fruits for major species and cultivars. Although 'typical' specimens have been chosen, each tree's potential variety cannot be illustrated, but *is* described. Many individual trees have also been portrayed but, as a tree's shape is particularly dependent on its environment, these must not be assumed to be characteristic. This book is intended as a portable companion to anyone looking at trees in the countryside, parks and gardens of Britain and Ireland.

Deciding which species to include in a tree guide is difficult not only because of the range of species and cultivars – at least 6000 – in Britain's collections as a whole, but also because the concept 'tree' is a vague one. 'Bird' is a precise scientific term, but 'trees' occur among groups of plants all across the evolutionary spectrum from ferns to palms. The beech family (Fagaceae) is a group of plants across which similarities of flower-structure suggest a close relationship; all are trees. But the rose family (Rosaceae) contains a jumble of herbs, shrubs and trees (such as cherries), while the figwort family (Scrophulariaceae) consists of herbs such as foxglove and just one genus of tree: the foxglove trees or *Paulownia*. For the purposes of this book, a 'tree' is any plant which commonly grows to 3 m on a single stem at least 20 cm thick. The rules have been bent to include some common natives (Dogwood, Alder Buckthorn) which very seldom do this, and to exclude many *Rhododendron*, *Pieris* etc., which eventually can, but of which old examples are rare.

Our aim has been to keep to the spirit of the late Alan Mitchell's *A Field Guide to the Tree of Britain and Northern Europe* (1974), and I have been more than happy to take the opportunity to borrow several descriptions that could not be bettered. Emphasis has continued, where possible, to be placed on those features (bark, habit, shoots or leaves) which best assist field-identification year-round, rather than the flowers and fruit which display more absolute distinguishing characteristics but are often unavailable. Many exotic trees have also been widely cultivated to date in a limited range of clones, and this has allowed us to treat as diagnostic some features of bark and habit which may become less reliable in the wild (or among the wild-collected specimens in major arboreta). Botanical terminology has been kept to a bare minimum.

Looking at Trees

Tree-watching is an engrossing but under-subscribed hobby. Although trees are big organisms, they are easily overlooked, especially perhaps by eyes tuned to the movements of tiny birds or used to scanning the ground for rare herbs. Their impact on our surroundings is as overwhelming as that of landforms or weather, and is likely to be taken for granted in the same way: only when we look at old photographs or confront the effects of a freak storm do we realize with a jolt how much of a difference the growth, the loss and the variety of trees can make. It was discovered only in 1989 (by the eminent field botanist Dr Francis Rose) that one of the trees growing – in towering, wild groves – on the scarps of the South Downs is the very rare Broad-leaved Lime; contemporaneously, ancient True Services, identified on cliffs near Cardiff, added a new species to the list of British natives. An entirely new species – *Zelkova sicula* – was discovered growing as a population of 200 trees in southeast Sicily in 1991. Everywhere, too, are long-forgotten arboreta, where knowledge of the site's origins has long been lost but the rare trees persist, standing incongruously in suburban verges, in the waste ground behind an industrial estate, or in an overgrown spinney. Most cemeteries and town parks have at least one tree rarer than passers-by will in general realize, and the definition of a tree in this book as 'rare' should not tempt you to assume a misidentification.

Learning to recognize trees is an open-ended process. The first steps are hardest: once you can confidently distinguish a hornbeam from a beech, you can begin to notice similar trees which are clearly something different, and you will also be able to guess whether the tree is a rarer hornbeam or represents a new group. Human brains habitually skip the unfamiliar: it can be salutary to spend time studying every tree in, say, a churchyard – even the small ones in the corner shrubberies – as the process may lead you to identify a species you have never consciously spotted before but which, once learnt, will crop up everywhere.

Winter is the hardest time to identify trees, as broadleaves with neither flowers nor foliage offer few clues, and rarities are bound to be missed. Autumn is a good study period, as foliage tints highlight differences and leaves, out of reach in summer, can be picked off the ground. (Leaves which decay slowly, such as the Wild Service's, can blow large distances during winter and examples may have to be tracked down by following trails of foliage.) Spring, when winter buds have burst but leaves are not expanded, is a tricky time but the only window for identifying some flowering trees.

Binoculars are useful tools (leaves may be just too high up for details of toothing to be seen from the ground; most silver firs carry cones only near their tops). A hand-lens (×10) can also reveal minute tell-tale hairs or glands on leaves or shoots.

Labelled arboreta are the best places to learn the less common trees, though it must be borne in mind that labels on the trees in many town parks, for example, are frequently incorrect.

Emphasis has been placed in this book on the most concrete or fail-safe ways of differentiating trees: are the buds, for instance, alternately spaced or in opposite pairs; do the leaves have hairs underneath? This yes-or-no methodology lends itself to description and depiction, but quite quickly you will instead be able to stand at a cemetery gate or look out of a train window and identify every tree you can see simply from subtle variations in colour, texture and habit.

A mature Wych Elm in winter

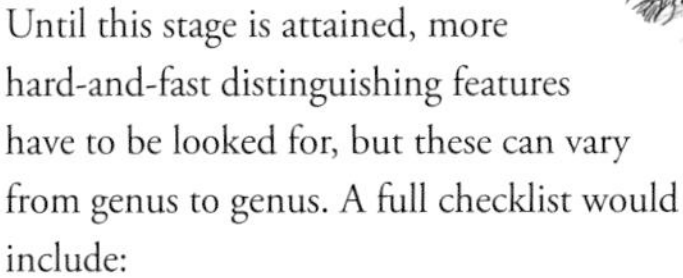

Until this stage is attained, more hard-and-fast distinguishing features have to be looked for, but these can vary from genus to genus. A full checklist would include:

- **SHAPE**: Do the branches weep, or rise steeply, or spread in plates? Does the tree look as if it will become large?
- **BARK**: Is it rough or smooth (at what age/diameter)? What colour is it? (In clean or moist air the bark may be covered by moss, lichens or green and orange algae.) Is there a graft? (a horizontal discontinuity, showing that a nursery has 'worked' a cutting of a clone, or of a species hard to grow from seed, on to an easily propagated rootstock. A graft suggests a rarity or a nameable cultivar, though by no means all such trees will be grafted. Sometimes a loop of wire grown into the trunk can create a graft-like scar.)
- **SHOOTS**: Are they hairy, or grooved, or warty? What colour are they? (The hairs may be tiny and transparent. The easiest way to spot them is to hold a specimen against a dark background next to a light-source; they will catch the light and halo the shoot.) Is there a distinct smell if you scratch the wood? Do the leaves/leaf-buds come in opposite pairs, or one by one?
- **BUDS**: What shape are they? Are they hairy? Have they any protective scales, and of what colour?
- **LEAVES**: How big are they? (Resist the temptation to choose the biggest leaf you can find: measure a well-grown, average one.) What shape? Are the margins lobed or toothed? Are there any hairs – above, below, along the veins, on the stalk, or in tufts under the main vein-joints? The underside of the leaf holds more clues than the top: what colour is it? How conspicuous are the veins?
- **FLOWERS AND FRUIT**: Their structural range and evolutionary stability make these the most useful parts of the tree for identification and the basis for the differentiation of scientific families and genera, though they will often not be available in the field.

Further Reading

The learning process does not end when you are familiar with the trees in this book. Works which can take you further include:

Trees and Shrubs Hardy in the British Isles by W. J. Bean (Murray; 8th Edition 1973–1989).
Manual of Cultivated Broad-leaved Trees and Shrubs and *Manual of Cultivated Conifers* by Gerd Krussmann (English translations 1984ff).
New Trees: Recent Introductions to Cultivation by John Grimshaw and Ross Bayton (Kew Publishing, 2009)

British and Irish Native Trees

Because of extinctions caused by successive Ice Ages, and the brief window between the end of the last Ice Age and the area's isolation by the flooding of the North Sea, the native tree flora of Britain and Ireland is singularly impoverished. Excluding hybrids, varieties and unashamed bushes, the 80 species and microspecies that seem most likely to have got here without human agency are:

Field Maple *Acer campestre* (England)
Common Alder *Alnus glutinosa*
Strawberry Tree *Arbutus unedo* (SW Ireland; local)
Silver Birch *Betula pendula*
Downy Birch *Betula pubescens*
Box *Buxus sempervirens* (SE England; local)
Hornbeam *Carpinus betulus* (SE England)
Dogwood *Cornus sanguinea* (England, Wales and Ireland)
Hazel *Corylus avellana*
Midland Thorn *Crataegus laevigata* (S England)
Hawthorn *Crataegus monogyna*
Spindle *Euonymus europaeus*
Beech *Fagus sylvatica* (S England, S Wales)
Alder Buckthorn *Frangula alnus* (England, Wales and Ireland)
Common Ash *Fraxinus excelsior*
Sea Buckthorn *Hippophae rhamnoides* (E coasts of England and of S Scotland)
Holly *Ilex aquifolium*
Juniper *Juniperus communis* (local)
Wild Crab *Malus sylvestris*
Scots Pine *Pinus sylvestris* (Scottish Highlands; naturalized elsewhere)
White Poplar *Populus alba*
Black Poplar *Populus nigra* ssp. *betulifolia* (England and Wales; local)
Aspen *Populus tremula*
Wild Cherry *Prunus avium*
Bird Cherry *Prunus padus* (N Britain)
Blackthorn *Prunus spinosa*
Plymouth Pear *Pyrus cordata* (SW England; very rare)
Wild Pear *Pyrus communis* ssp. *communis* (England; very rare)
Sessile Oak *Quercus petraea*
English Oak *Quercus robur*
Purging Buckthorn *Rhamnus cathartica* (England, Wales and Ireland)
White Willow *Salix alba*
Goat Willow *Salix caprea*
Grey Sallow *Salix cinerea* ssp. *oleifolia*
Crack Willow *Salix × fragilis*
Bay Willow *Salix pentandra* (N Britain and Ireland only)
Purple Osier *Salix purpurea* (local)
Almond Willow *Salix triandra* (local)

Common Osier *Salix viminalis*
Elder *Sambucus nigra*
Sorbus admonitor (N Devon)
Sorbus anglica (SW England, Wales and County Kerry; local)
Whitebeam *Sorbus aria* (S England; Ireland)
Arran Whitebeam *Sorbus arranensis* (N Arran)
Rowan *Sorbus aucuparia*
Bristol Service *Sorbus bristoliensis* (Avon Gorge)
Sorbus cambrensis (Brecon Beacons)
Sorbus cheddarensis (Cheddar Gorge)
Sorbus cuneifolia (Eglwyseg Mountain, Denbighshire)
French Hales *Sorbus devoniensis* (SW England and SE Ireland; local)
True Service *Sorbus domestica* (Wales; very rare)
Sorbus eminens (Wye Valley and Avon Gorge)
Sorbus eminentoides (Cheddar Gorge)
Sorbus hibernica (central Ireland; local)
Sorbus lancastriensis (S Cumbria)
Sorbus leighensis (Avon Gorge)
Sorbus leptophylla (Brecon Beacons)
Sorbus leyana (Taff Valley, Glamorgan)
Sorbus margaretae (N Devon)
Sorbus minima (Brecon Beacons)
Sorbus porrigentiformis (N Devon, Mendips and S Wales; local)
Arran Service *Sorbus pseudofennica* (N Arran)
Sorbus pseudomeinichii (N Arran)
Sorbus robertsonii (Avon Gorge)
Cliff Whitebeam *Sorbus rupicola* (limestone uplands; local)
Sorbus rupicoloides (Cheddar Gorge)
Sorbus scannelliana (Killarney, Ireland)
Sorbus stenophylla (Llanthony valley, Powys)
Sorbus stirtoniana (Craig Breiddon, Powys)
Exmoor Service *Sorbus subcuneata* (Exmoor)
Wild Service *Sorbus torminalis* (England and Wales)
Sorbus vexans (N Devon)
Sorbus wilmottiana (Avon Gorge)
Sorbus whiteana (Avon Gorge).
Common Yew *Taxus baccata*
Small-leaved Lime *Tilia cordata* (England and Wales; local)
Broad-leaved Lime *Tilia platyphyllos* (England and Wales; very local)
Wych Elm *Ulmus glabra*
European White Elm *Ulmus laevis* (very rare)
Field Elm *Ulmus minor* (in England, in at least some clones)

WELL-NATURALIZED, LOCALLY AT LEAST, ARE:

Norway Maple *Acer platanoides*
Sycamore *Acer pseudoplatanus*
Horse Chestnut *Aesculus hippocastanum*
Snowy Mespil *Amelanchier lamarckii*
Sweet Chestnut *Castanea sativa*
Common Apple *Malus pumila*
Medlar *Mespilus germanica*
Grey Poplar *Populus canescens*
Myrobalan Plum *Prunus cerasifera*
Sour Cherry *Prunus cerasus*
Common Plum *Prunus domestica*
Bullace *Prunus insititia*
Cherry Laurel *Prunus laurocerasus*
Portugal Laurel *Prunus lusitanica*
St Lucie Cherry *Prunus mahaleb*
Black Cherry *Prunus serotina*
Common Pear *Pyrus communis* ssp. *sativa*
Turkey Oak *Quercus cerris*
Holm Oak *Quercus ilex*
False Acacia *Robinia pseudoacacia*
Sorbus croceocarpa
Tamarisk *Tamarisk gallica* and other species

Plants' scientific names are always worth learning, as common names can vary from region to region. 'Sycamore' in the United States is a species of *Platanus* ('plane'), while in Scotland 'plane' can be *Acer pseudoplatanus* ('Sycamore'); 'Siberian Crab', used in this guide for *Malus baccata*, is used by some gardeners for *M.* × *robusta*.

Customarily italicized, a scientific name has two main parts: the generic name ('surname'), followed by the specific name ('forename'): *Quercus* (never *quercus*) is the name for the group of related trees commonly called 'oaks'. *Quercus robur* (never *Robur*) is one distinct species, the English Oak. The fuller form of the name is *Quercus robur* L., the 'L.' denoting the botanical authority who published this name: the eighteenth-century Swedish botanist, Carl von Linné, who wrote as 'Linnaeus'. The convention antedates Darwin's theories of evolution: any nomenclature is ultimately a convenient but artificial way for us to parcel up the constantly evolving continuum of the genes of all life.

Obvious sexual hybrids between species are denoted with an '×': *Quercus* × *rosacea* is the hybrid of *Quercus robur* and *Quercus petraea* (Sessile Oak), but this entity can just as legitimately be called *Quercus petraea* × *robur*. Fertile, stable hybrids (e.g. *Sorbus hybrida*) generally lose their '×'. If two trees customarily allocated to different genera hybridize, a new generic name is needed and the '×' precedes this: × *Crataemespilus grandiflora* is the hybrid of *Crataegus laevigata* (Midland Hawthorn) and *Mespilus germanica* (Medlar). Sometimes hybrids occur when genetic material, rather than mixing sexually, fuses in a natural grafting process, creating a 'chimaera'; for these trees, a '+' is used. *Aesculus* + *dallimorei* (Dallimore's Chestnut) is a fusion of tissues of Common Horse Chestnut (*Aesculus hippocastanum*) and Yellow Buckeye (*Aesculus flava*), with cells of both entities co-existing in a single plant. + *Laburnocytisus adamii* is an intergeneric chimaera, Common Laburnum (*Laburnum anagyroides*) having fused with a purple broom (*Cytisus purpureus*). Some genera (e.g. *Sorbus*) are partly divided not into species but into microspecies, whose individuals are self-fertile and go on producing seedlings with very little genetic diversity.

Quercus x *hispanica* 'William Lucombe'

A third botanical name is sometimes used to distinguish different forms of a species – either subspecies (ssp.) or variety (var.); variety suggests a difference in appearance rather than habitat. Corsican Pine is *Pinus nigra* ssp. *laricio*, Austrian Pine *Pinus nigra* ssp. *nigra*. A third category, forma ('f.'; plural formae), may be used to describe a population of 'sports': Sycamores with purple-backed leaves – often found in the wild – are *Acer pseudoplatanus* f. *purpureum*.

A 'cultivar' is, strictly, one particular clone or asexually propagated individual plant, and its name is never italicized because the rules of scientific nomenclature no longer obtain. Instead, the first letter is capitalized and the word is enclosed in single quotation marks *or* preceded by 'cv': *Malus pumila* 'Granny Smith' or *Malus pumila* cv. Granny Smith. Nurseries naming their cultivars traditionally used pseudo-scientific names (e.g. 'Pendula' for weeping trees); since 1959 names have been required to be in the vernacular (e.g. *Gleditsia triacanthos* 'Sunburst'). In practice, cultivar names often serve to cover 'variants' (groups of clones with similar characteristics): *Populus nigra* 'Italica' is used for Lombardy Poplars of various inherited shapes.

Corsican Pine, *Pinus nigra* ssp. *laricio*

The allocation of families and genera is an art with no written laws, but the rules for naming species are precise: the correct name is the earliest one to have been published in a reputable journal which unambiguously describes an understood species. A plant's specific name must not be the same as its generic name, though an animal's can be (e.g. *Troglodytes troglodytes*, the Wren). Names change for three reasons: a new description, published earlier, may be rediscovered; an existing name may be deemed to embrace trees from another species; or new ideas on a tree's family relationship may come into fashion, prompting a different generic name (plus the adoption of the next-oldest species name if the previous one is already in use in the new genus for a different tree).

Spellings can be important: *Acer maximowiczii* is a rare Snake-bark Maple; *Acer maximowiczianum* is the very dissimilar Nikko Maple. *Acer pensylvanicum* ('from Pennsylvania') is spelled like this because of a misprint in Linnaeus' original description, though his *Stewartia* (honouring John Stuart, Earl of Bute) is often 'corrected' to *Stuartia*. Specific names should inflect (like Latin adjectives) to agree with the gender of the genus name – but as botanists are seldom classical scholars there is frequently an unresolved confusion between, for example, '*europaeus*' and '*europaea*'.

Common Ash,
Fraxinus excelsior

THE TREES

MAIDENHAIR TREE

The Maidenhair Tree is one of the most ancient and singular trees we can grow. The Ginkgoaceae family, dominant in the Mesozoic Era, has only this one surviving species.

Maidenhair Tree

Ginkgo biloba

Brought from temple gardens in Chekiang Province, China, in 1758; now endangered in the wild. Quite frequent in areas with warmer summers.

APPEARANCE **Shape** Haphazard, and spiky with short 'spur' shoots; spire-shaped with few small, horizontal limbs from a slightly leaning trunk, or on boles narrowly diverging from low down. To 28 m. **Bark** Grey-brown, corky then craggy, its exposed growth-rings sometimes creating a beautiful lace-like silver patterning; oldest specimens – still healthy – developing the 'chi-chi' or 'breasts' which characterize ancient Chinese trees. **Shoots** Shiny, grey. **Buds** Fat but sharp, green/red. **Leaves** *Fan-shaped*, 9 × 7 cm; alternately up strong shoots and clustered on short 'spurs'. Fresh green; rich gold in late autumn. The marginal notches are variable and tend to be deeper on young plants. **Flowers** Trees either male or female. Male catkins yellow, 2–4 cm, in bunches in late spring; scarcely seen in Britain. Female flowers green 4 mm knobs, paired on 4 cm stalks; plum-like fruits ripen yellow in autumn (given enough summer warmth) and smell unpleasant. Grafts from male trees are consequently preferred as garden plants. The white nut in the fruit is, however, a delicacy in China.

COMPARE In winter, Golden Larch: a more symmetrical, broad-conic crown and square-cracked bark. Some specimens of Common Pear (p.177): similar spiky crown, dense with 'spur' shoots, but blackish bark.

VARIANTS 'Variegata' is very rare, as is the low, weeping 'Pendula'. Straight, narrow forms have been selected, especially in USA where they are invaluable as heat-tolerant street trees ('Sentry'; 'Fastigiata'). The best clones have tight crowns on vertical twigs and recall Lombardy Poplar, but very slender trees are part of the natural variation.

old tree

fastigiate form

YEW

Yews (six similar species) are ancient trees with berry-like fruits – red 'arils'. (Family: Taxaceae.)

Things to Look for: Yews to Podocarps

- Bark
- Leaves: How big? What colours are they underneath? Are they spine-tipped? Do they curve (up, down)?

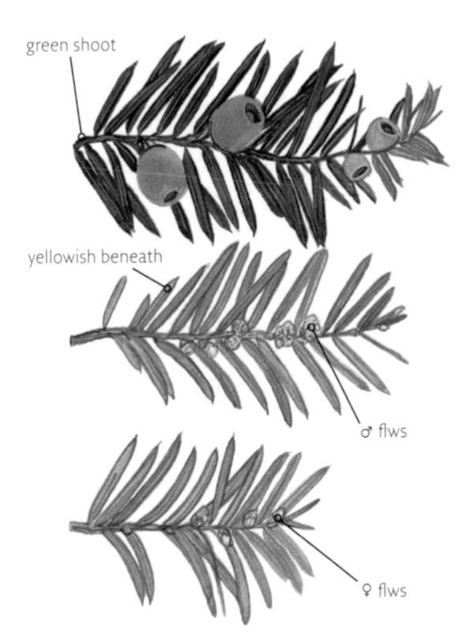

Common Yew

Taxus baccata

Europe, N to southern Scandinavia and E to Iran, and including Britain and Ireland; N Africa. Grows singly in woods and hedgerows on mineral soils, with some dense groves on limestone scarps, but long eradicated from many parts because it is toxic to livestock; abundantly planted in parks, shrubberies and gardens. One of the world's longest-lived trees: over 2000 years old in many English, Welsh and N French churchyards.

APPEARANCE Shape Often soon broad, but generally with a pointed tip; to 25 m. Heavy, blackish presence. **Bark** Shallowly scaling; grey, purplish and reddish – can be exquisitely colourful. **Shoots** *Green for three years*, with 2 mm *round* green buds.

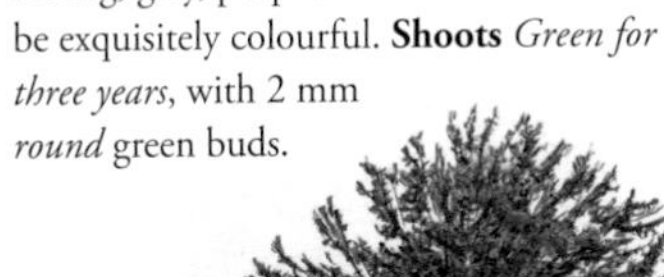

Leaves 30 × 3 mm, downward-arching, in one flattened rank on either side of the side-shoots but spiralling all round strong/erect growths; abruptly sharp-pointed but not rigid; dull *yellowish* grey-green beneath. Highly toxic: do not crush them to release scent. **Flowers** Nearly always dioecious. Male flowers whiten the crown before shedding clouds of pollen in early spring. **Fruit** Arils ripen in early autumn, cerise. The flesh wrapping the lethal seed is sweet.

COMPARE Coast Redwood (p.40): disparate scale-leaves clasping its shoots. Among trees with shoots green for some years, the yews' slightly yellowish underleaves are diagnostic.

VARIANTS 'Adpressa' (1828 or 1838) has *half-length*, full-width leaves; dark blue-grey and matt *en masse*. Female; rare. 'Adpressa Variegata' ('Adpressa Aurea'; 1866) has yellow-edged, half-length leaves.

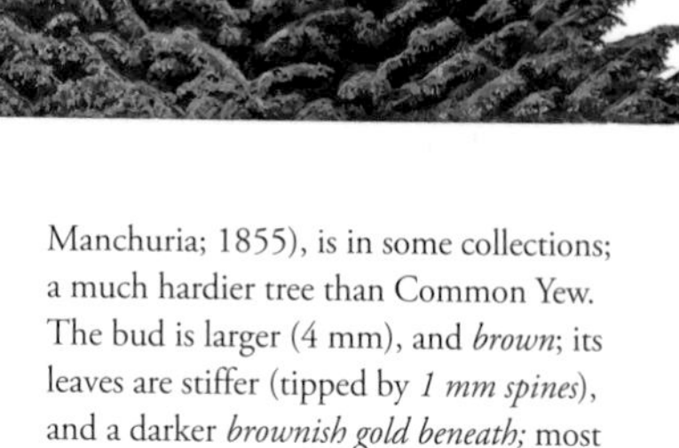

WESTFELTON YEW

Golden Yew, f. *aurea*, has yellow-edged younger leaves; frequent in gardens and, rarely, found wild.

Westfelton Yew, 'Dovastoniana', arose in Shropshire by 1776 (where the original still thrives): its foliage *cascades from wide very low limbs*. Female; very occasional. 'Dovastonii Aurea' has yellow-edged younger leaves; rare and low.

'Lutea' ('Fructu-Luteo'; 1817) has *yellow* fruit; very rare but very striking.

Irish Yew, 'Fastigiata' ('Stricta', 'Hibernica'), is abundant; *erect* (narrowest in high-rainfall areas; cf. Hicks' Yew, below); blackish leaves all round *all* shoots; normally female. One of two original seedlings (transplanted by 1780) still grows at Florencecourt, Co. Fermanagh. 'Fastigiata Aurea' (male: yellow-edged leaves) is frequent; the female equivalent 'Standishii' (1908) is very bright and dense but slow and occasional.

OTHER TREES Hicks' Yew, *T.* × *media* 'Hicksii', a rare clone of a hybrid with Japanese Yew (1900; to 9 m so far), resembles Irish Yew. A female plant, with large fruits; leaves *yellower* beneath, tipped by a short, quite soft *spine*; crown blobby and fuzzy, less crisply turreted. (Fastigiate Plum Yew, has bigger leaves, whiter beneath.)

Japanese Yew, *T. cuspidata* (Japan, Korea, Manchuria; 1855), is in some collections; a much hardier tree than Common Yew. The bud is larger (4 mm), and *brown*; its leaves are stiffer (tipped by *1 mm spines*), and a darker *brownish gold beneath;* most *arch almost to vertical*; females carry massed, *clustered* fruits.

Chinese Yew, *T. mairei*, has been grown (usually as *T. celebica*) in a few collections since 1908 – a sad bush. Slender, *yellowish, sparse* leaves curl backwards on either side of a shoot which turns *brown after one year*; its fruits *stay green*.

IRISH YEW

MONKEY PUZZLE

Araucariaceae is a small 'living fossil' family.

Monkey Puzzle

Araucaria araucana

(Chile Pine; *A. imbricata*) When the Scottish plant-hunter Archibald Menzies was given unfamiliar nuts for dessert during a banquet in Valparaiso in 1792, he is supposed to have slipped five into his pocket, germinating them on the voyage home. The 'monkey-puzzler' became a sensation and there is now believed to be more genetic diversity in European gardens than in the vulnerable remaining Andean stands (all on the slopes of dormant volcanoes). Occasional in gardens of all sizes, living longest in high-rainfall areas; sometimes seeding. Its instant recognizability makes it seem commoner than it is.

APPEARANCE Shape Trees vary in breadth, but have a straight trunk (very rarely forked) designed to carry the foliage quickly out of browsing dinosaurs' reach. To 30 m. The branch-whorls, whose scars are conspicuous up the trunk, are not produced annually: growth can stop for the winter at any stage and the average whorl represents some 18 months' growing. **Leaves** 4 × 2 cm; leathery, spined. Felled or blown trees coppice, and the regrowth can be sinuous, with smaller leaves. **Flowers** Nearly always dioecious. **Female cones** 15 cm across; they ripen in two years and – fortunately – break up on the tree. The nuts (rather like Brazil nuts) are best roasted.

OTHER TREES Parana Pine, *A. angustifolia*, an important forestry tree from SE Brazil/Argentina, is scarcely hardy in Britain but grows in one or two collections, with slenderer, softer, sparser leaves.

CYPRESSES

The family Cupressaceae embraces most of the trees known as 'cypresses'. The minute, scale-like adult leaves make ferny sprays; juvenile leaves, persistent in some false cypress cultivars and in junipers, are short, paired, spreading needles – finer and straighter than those of Japanese Red Cedar (p.39). Cypresses can carry '2D' (fern-like) or '3D' (plume-like) sprays (usually juvenile) – but keep the type's distinctive aroma.

THINGS TO LOOK FOR: ALL CYPRESSES

- Shape: How weeping?
- Leaves: Are foliage sprays 2D or 3D? Aroma? How glossy? White bands/spots (especially beneath)? Tip (how spreading? how sharp?).
- Cones or berries: How big? What colour?

KEY SPECIES ✪

Incense Cedar (below): 2D sprays (long side-scales). **Western Red Cedar** (p.22): 2D sprays (broad, smooth). **Lawson Cypress** (p.25): 2D sprays (finer, almost smooth). **'Squarrosa' Cypress** (p.29): almost 3D sprays of soft juvenile leaves. **Leylandii** (p.31): almost 3D sprays (long-pointed scales). **Monterey Cypress** (p.32): fully 3D sprays (very smooth). **Common Juniper** (p.36): 3D sprays of sharp juvenile leaves. **Chinese Juniper** (p.37): fully 3D sprays (round smooth threads plus some sharp juvenile leaves).

Incense Cedar ✪

Calocedrus decurrens

(*Libocedrus decurrens*) Mid-Oregon to S California. 1853. Occasional in larger older gardens, and increasingly fashionable. Resistant to honey-fungus and *Phytophthora* root-rot.

APPEARANCE Shape In Britain a *dense exclamation-mark* (to 40 m). Wild trees, curiously, are open and level-branched. Irish specimens are broader, with erect branching, and Continental ones often narrow, but with more spreading branches. **Bark** Rich red-brown; long, curling or spongy plates. **Leaves** In rising sprays. Side ones with tiny sharp points (fractionally incurved so the sprays are not prickly); longer than most cypresses', giving the broad (3 mm) shoots a *stretched* appearance. Rich matt green; *no white*

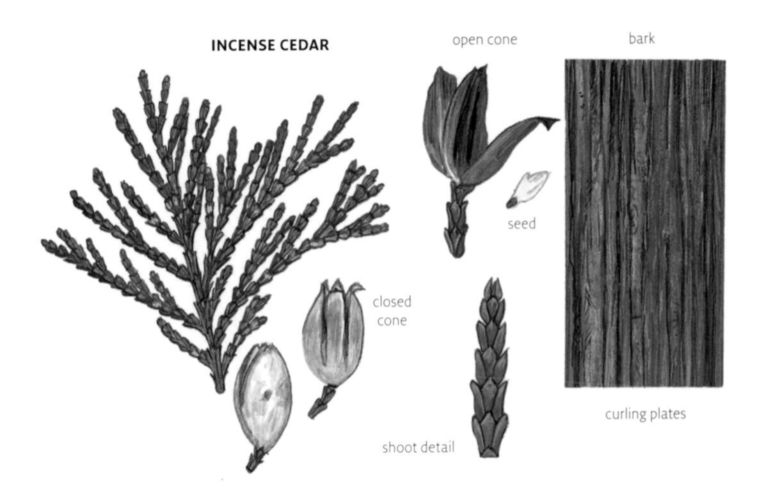

marks beneath. Shoe-polish aroma. **Cones** 25 mm, with six scales, yellow then rufous; like lines of mantelpiece vases.
COMPARE False cypresses (pp.25–31): finer sprays of shorter leaves. Nootka Cypress (p.30): equally long leaves, hanging foliage. Lawson Cypress 'Erecta' (p.26): similar at a distance (more egg-shaped). Oriental Thuja (p.24) and Western Red Cedar 'Fastigiata' (p.23) also have erect sprays.
VARIANTS 'Aureovariegata' (1904), slower growing but shaped as the type, has *blotches of yellow foliage*.

Chilean Incense Cedar
Austrocedrus chilensis

(*Libocedrus chilensis*) S central Andes. 1847. In some collections in milder areas.
APPEARANCE **Shape** Columnar, but *only to 15 m*; a soft sage-green. **Bark** Greyer than Incense Cedar's; *small* scales. **Leaves** Fungous scent; *broad white marks beneath, and often on top*. The side-leaves' incurving tips project, but are *blunt* (cf. Hinoki Cypress, p.27). **Cones** Only four scales (rarely seen in N Europe).

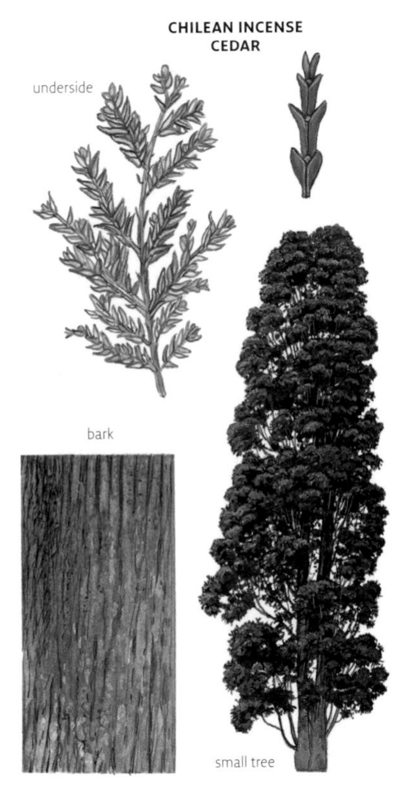

Hiba

Thujopsis dolabrata

An important forestry tree in Japan. 1853. Occasional: most frequent in wet areas; sometimes in small gardens for its lavish foliage.

APPEARANCE Shape A broad spire, to 25 m, or a ring of layered stems. **Bark** Red/grey; stringy strips. **Leaves** In 2D sprays *5 mm wide*, lizard-like; convex and *very glossy green* above, a big *dead white* curved streak beneath.

COMPARE Chilean Incense Cedar (p.21). Korean Thuja: much slenderer shoots.

VARIANTS Many were planted as 'Variegata', with a sprinkling of cream shoots; soon few spots remain.

'Aurea' (1866) is very rare but quite a bright gold.

While in the same family as cypresses, trees in the *Thuja* genus have 2D sprays of neat, rather flattened, richly and sweetly aromatic foliage-threads; the minute leaf-points are incurved so that the shoot feels smooth. The oval cones stand above the branches. Sometimes called Arbor-vitae ('tree of life'); the foliage of Eastern White Cedar (rich in vitamin C) was infused to combat scurvy (the leaves in bulk are poisonous/allergenic). (Family: Cupressaceae.)

Western Red Cedar

Thuja plicata

(*T. lobbii; T. gigantea*) A giant tree from NW North America, growing on windswept hillsides where its tops are repeatedly blown out but the basal trunk can survive for several millennia. 1853. Abundant in parks, gardens and hedges (though occasional in dry/polluted parts); some forestry plantations; self-seeding in old quarries, neglected mortar, etc.

APPEARANCE Shape When growing well, a tidy spire with a billowing base and erect leader, like a rocket taking off; sometimes forked or layering. *Bigger* than any similar cypress: 45 m so far. Broader, sparse or broken-topped in dry conditions. **Bark** Dark red-brown; soft, stringy ridges (often greyer and harder in plantations). **Leaves** In luxuriant, rather *glossy* drooping sprays, the sheen accentuating their neat design; a fresh deep olive-green (slightly bronzing in winter), with dull yellowish/white-green streaks beneath. *Strong, sweet pineapple/Philadelphus scent* fills the air on warm days.

COMPARE Eastern White Cedar and Japanese Thuja (p.23). Lawson Cypress

(p.25): finer, duller shoots, rather sourly parsley-scented. Nootka Cypress (p.30) and Leylandii (p.31): shoots the same width, but dull and dark with spreading leaf-points and a sourer smell. Hiba (opposite): much bigger, glossier leaves.

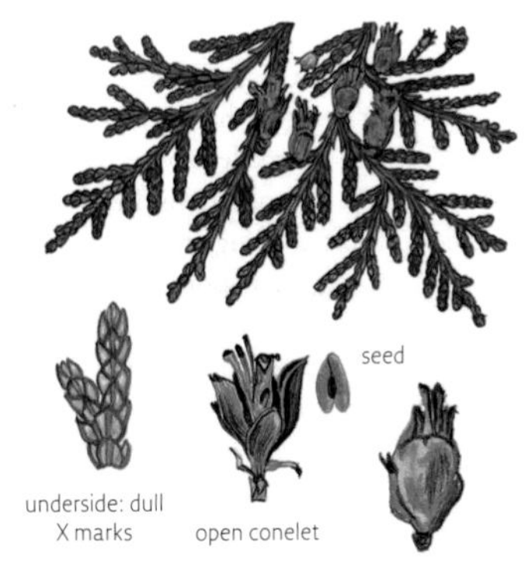

VARIANTS 'Zebrina' (1868) is a bright *pale yellow* abundant garden tree, the colour formed by *zebra-stripes* of white-gold, green-gold and green across each spray. More than one clone: sometimes a tall spire (to 28 m); more often ('Irish Gold') slow and dumpy but very bright.

'Aurea' (still rare) has big, dishevelled, uniformly dark gold sprays. The very rare 'Semperaurescens' is greener, but neat and glossy. 'Wintergold' (rather rare) has scraggily rising, gold-tipped sprays.

'Fastigiata' (1867), occasional in hedges and parks, has dark sprays held *well above the horizontal*, and a *narrow*, tight spire shape (though the lower branches frequently layer).

Japanese Thuja

Thuja standishii

Japan. 1860. Very occasional.

APPEARANCE Shape Easily taken for a dumpy (to 24 m) yet imperturbably healthy Western Red Cedar, but subtly and prettily distinctive. **Bark** Develops harder (even shiny) ridges, some *crimson-pink*. **Leaves** *Matt*, the sprays *rougher* above; fine, brighter grey-white streaks underneath; greyer at first – each short, drooping, rather distant spray *neatly tiling the crown*.

OTHER TREES Hybrids with Western Red Cedar (intermediate) are growing fast in some collections.

Oriental Thuja

Platycladus orientalis

(Biota; *Thuja orientalis*) China – where known only from gardens. 1752. An oddball conifer, appreciating hot, dry sites and a rich/chalky soil; very hardy, yet thriving in the tropics. Quite frequent in lowland parks, old gardens, and especially country churchyards; rarely naturalizing.

APPEARANCE **Shape** A *balloon*, to 15 m: slightly sinuous stems from a short, grey-brown trunk. **Leaves** *Scentless;* in rather vertical 2D plates, *the same colour each side*; shoots duller and finer than thujas'.

COMPARE Lawson Cypress 'Erecta' (p.26): a bigger, parsley-scented tree with minutely pointed side-leaves.

VARIANTS 'Elegantissima' has foliage-plates brightly yellow-tipped in summer, duller gold-green in winter. Frequent in rockeries and churchyards, to 10 m. Crisper habit than Eastern White Cedar 'Ellwangeriana Aurea' – like a giant coral.

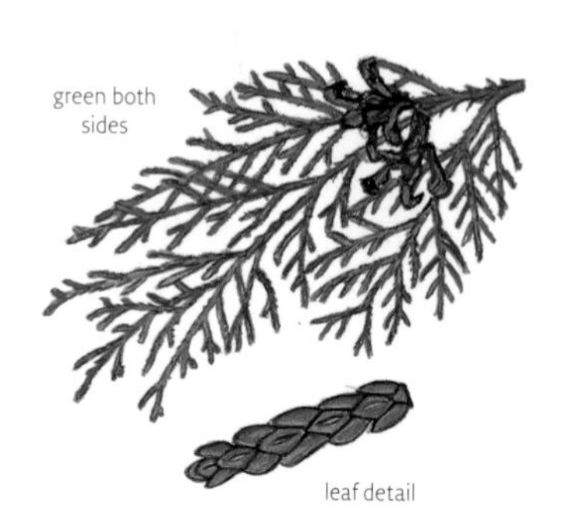

Lawson Cypress ✪ ◐
Chamaecyparis lawsoniana

(Port Orford Cedar) A few stands in the Klamath and Siskiyou Mountains (Oregon/California), now threatened by fungal attack. 1854. Very abundant in the UK, in parks, gardens, hedges. Used in forestry to 'nurse' beech plantations; naturalizing on sandy banks etc.

APPEARANCE Shape Densely columnar (to 42 m so far); open and broken in drier sites; bole often much forked. *Drooping leader* and nodding sprays, hanging like rags on some old trees. **Bark** Reddish and purplish, *spongy*; harder-plated with age but *never spirally ridged*. **Leaves** Matt threads 1.8 mm wide; side-leaves *minutely* free-pointed but not prickly, and grey-white beneath along the joints; particularly prominent *translucent*

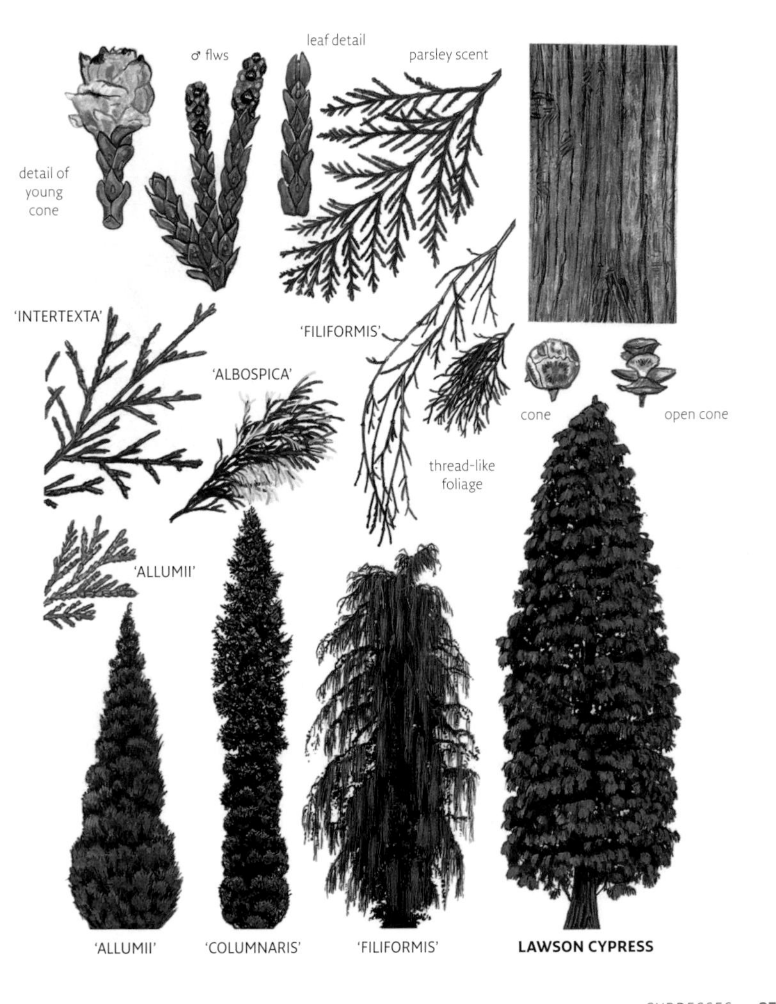

'ERECTA VIRIDIS'

'GREEN SPIRE'

'GRAYSWOOD PILLAR'

'INTERTEXTA'

gland in the centre of the top-leaves. Rather sour, *parsley* scent. **Cones** Abundant, pea-sized, ripening in one year.

COMPARE Nootka Cypress (p.30): longer, longer-pointed side-leaves. Sawara Cypress (p.28) and Taiwan Cypress (p.30): side-leaves with more spreading points. Hinoki Cypress (opposite): blunt side-leaves. Thujas (pp.23–4): broader, smoother shoots.

VARIANTS A uniform tree in the wild, but after its arrival in Europe it immediately began to throw more 'sports' than any other species. Some are now the mainstay of small gardens. Those with aberrant leaves *still have the parsley scent*. With age, foliage colours tend to fade and habits become less pronounced.

Variegated forms include 'Albospica' (shoots brilliantly white-blotched/tipped), 'Versicolor' (gold-and-white speckling) and 'Albomaculata' (tall-growing; big ivory-yellow splodges). None are common.

Forms with erect foliage include 'Erecta Viridis' ('Erecta'), which grew from the first consignment of seed raised in Britain. Frequent, to 35 m: tightly *egg-shaped* (pointed at first; much broken with age); vivid green. Sprays rise steeply (their tips may nod); leaves slightly bigger than the type's but much shorter than Incense Cedar's (p.20); cf. Oriental Thuja (p.24). 'Kilmacurragh' (by 1951) makes a *narrow column*; still rare. 'Allumii' (1890; abundant in small gardens, but to 30 m) has largely edge-on, steely-grey sprays; 'Fraseri' (*c.*1893) makes a tighter, dull-green spire (cf. 'Youngii', below, and 'Green Spire'). 'Columnaris' (*c.*1940) is much narrower than 'Allumii', though with looser sprays at all angles. Fast-growing (to 22 m so far); its habit *abruptly degenerates* at 8 m. 'Grayswood Pillar' (by 1960) is marginally bluer. Both (?) are abundant. 'Green Spire' ('Green Pillar'; by 1947) has edge-on, *yellowish-green* sprays; *narrowly* spire-shaped; rare.

Seedlings can be very pendulous. Distinct weeping clones include 'Filiformis' (1878), spectacular but rare, with *thread-like streamers*, and 'Pendula Vera' (very rare), whose *smaller branches hang* as well as the sprays.

Forms with aberrant foliage include 'Intertexta' (1869), with big, distinctively *distantly branching,* drooping, lacy sprays of *thick shoots*. To 30 m; often forking; occasional in big gardens. 'Filifera' has distantly branching sprays of very *slender shoots*; very rare. 'Youngii' (by 1874) has *level, upcurling* sprays of *thick foliage*, bright green. A spire, to 25 m; rare.

Hinoki Cypress

Chamaecyparis obtusa

Mountains of central and S Japan – one of the country's major timber conifers (the 'Japanese Gateway' at Kew Gardens is made from it). The Japanese name means 'Tree of Fire' – wild stands ignite readily, perhaps even from the friction of branches rubbing in the wind. Introduced when Japan was opened to westerners in 1861; occasional in larger gardens. Almost always a happy, healthy tree in Britain and Ireland, as are its several popular cultivars.

APPEARANCE **Shape** Variably dense on a generally straight, single trunk to 25 m (sometimes multi-stemmed). The rich green foliage scarcely weeps and is often gathered into many horizontally projecting, densely sculpted branch-systems – *'Japanese-looking'* – between which the trunk remains visible.

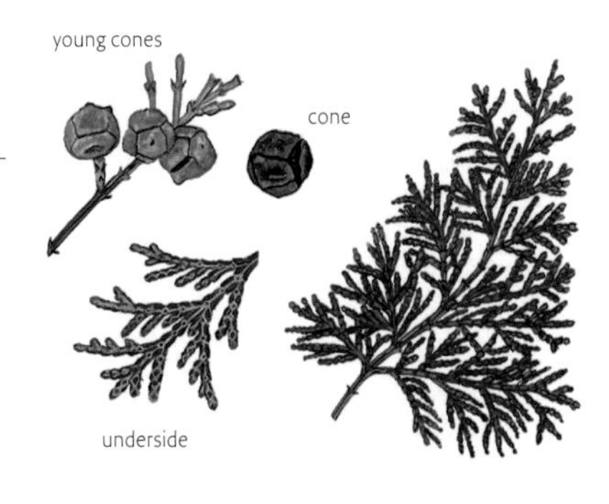

Bark Reddish, soft and rather stringy, or with harder vertical plates. **Leaves** In very neatly fanned sprays, matt yet rich green; side-leaves effectively *blunt*; each leaf with a crisp white base underneath. *Sweet*, Eucalyptus-like scent when crushed.

COMPARE Thujas (pp.23–4): broader shoots, none with such matt, horizontally fanned sprays. Chilean Incense Cedar (p.21): long-projecting, blunt side-leaves. True cypresses and junipers (pp.32–8): smooth '3D' shoots. The other false cypresses have minutely pointed side-leaves.

VARIANTS All have blunt scales and the same sweet smell. 'Crippsii' (1901) is one of the brightest and commonest of yellow cypresses, to 20 m so far. The pale gold

younger growth tipping each independent, horizontal branch-system makes these particularly distinct and distinguishes the tree at a distance.

'Tetragona Aurea' has *congested, blunt bright yellow turrets* and is very frequent. It was introduced in 1876 from Japan, where the green variety from which it must have mutated had long been lost to cultivation. Like many cypresses it is sold as a dwarf but grows a steady 10 cm a year, so that the oldest, 140 years old and still mostly thriving, are now 15 m tall. The dense '3D' foliage suggests a true cypress or juniper, but the yellow Monterey Cypresses (p.32) have longer, wispier, pointed turrets and Golden Chinese Juniper (p.38) is rather neatly conic, with odd patches of long, juvenile leaves.

Sawara Cypress

Chamaecyparis pisifera

Another giant cypress from the mountains of Japan. 1861. As frequent as Hinoki Cypress (p.27), but looking very different.

APPEARANCE Shape Rather *patchily sparse and open*, on a usually single but slightly sinuous trunk; particularly wretched in dry areas. **Bark** Softer and stringier than the other common false cypresses', red/grey. **Leaves** In rather meagre, mid-green sprays, slightly drooping; upper, lower and side-leaves *the same length*, with fine *spreading tips*; leaf-bases white underneath. Acrid, resinous aroma.

COMPARE Taiwan Cypress (p.30): very similar, but *no white markings* under its leaves. Lawson Cypress (p.25): more luxuriant sprays, the minute points of the side-leaves scarcely spreading. Nootka Cypress (p.30): longer, straighter side-leaves, not white beneath.

VARIANTS are numerous and highly diverse:

'Filifera' (1861, from Japan) has *thread-like hanging shoots* alternating with small bunched sprays; occasional. 'Filifera Aurea' (1889) is a small, *bright-yellow* variant, abundant in small gardens.

'Plumosa' (1861, from Japan) has prickly semi-juvenile foliage, the leaves with 3 mm points, in ascending and softly curling, slightly 3D sprays. Generally a leaning column on forking, easily shattered trunks.

The foliage, unlike the type's, is *very dense*, with dead leaves retained, so ideal for nesting birds. Frequent, especially in cemeteries, and healthier-looking than the species. To 25 m – but there are equally common semi-dwarf variants, such as 'Plumosa Compressa'. 'Plumosa Aurea' is also frequent: yellow-green, duller with age and reverting piecemeal to 'Plumosa'; the semi-dwarf 'Plumosa Aurea Compacta' is probably commoner.

'Squarrosa' ✪ (1843 – from Japan via Java) has effectively 3D sprays of long (6 mm), soft, paired juvenile leaves, bluish (from broad grey bands underneath) and contrasting prettily with its bright rufous bark. On a straight trunk/trunks, to 25 m; the massed, fuzzy sprays in horizontal branch-systems somewhat recall Hinoki Cypress. Frequent. (Common Juniper, p.36, and Meyer's Juniper, p.37, have rising sprays of much harsher needles, and are bushy; the Japanese Red Cedar 'Elegans', p.39, has soft needles carried alternately and twice as long, and a sprawling habit.) 'Boulevard' (1934), very abundant as a slender semi-dwarf (to 6 m), has slightly longer leaves in laxer, drooping pale sprays, the tips bronzing in winter. Live growth is sadly patchy with age. 'Squarrosa Sulphurea' is a frequent variant: pale creamy young foliage fades to the same blue-grey.

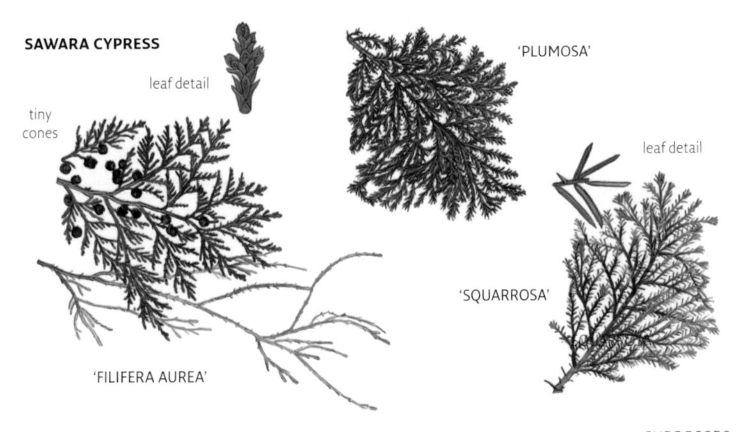

Taiwan Cypress

Chamaecyparis formosensis

(Formosan Cypress) A giant, threatened tree, perhaps to 3000 years old; in a few big gardens since 1910, but reaching only 18 m in the UK.

APPEARANCE Shape The branches typically curve up in big 'U's. **Leaves** Very like Sawara Cypress (p.28), but *lacking white marks* beneath. Foliage tends to bronze in winter. Aroma of *aniseed/seaweed.*

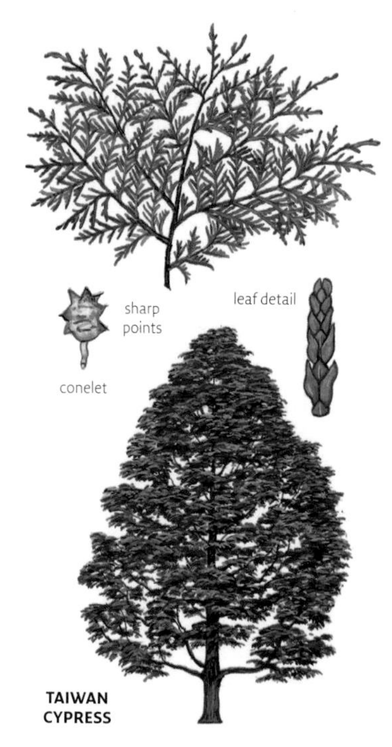

Nootka Cypress

Xanthocyparis nootkatensis

(*Chamaecyparis/Cupressus nootkatensis*) Perhaps the longest-lived tree in the cloud forests of NW North America, but unlikely to make a giant in Britain. 1854. Rather occasional, and happiest in high-rainfall areas. The 2D, sour-smelling sprays suggest a false cypress, but the cones take two years to mature.

APPEARANCE Shape Often neatly conic, to 30 m, though frequently on several trunks; a *solid surface* of close, *steeply hanging, dull, dark sprays,* with the interior clean

and open: *look up and you see the sky*. **Bark** Stringy grey-red ridges. **Leaves** Threads broad, like Western Red Cedar's (p.22), but matt; the straight, 3 mm side-leaves' slightly spreading points make the sprays *harsh if rubbed backwards; no white marks* on the yellowish undersides (cf. Incense Cedar, p.20, and Taiwan Cypress, opposite). Heavy, oily scent (like Stinking Mayweed). **Cones** *9 mm*, purple-green then red-brown in their second year.

COMPARE : Leylandii (below): *less pendulous*, imperfectly flattened sprays; columnar crown *dense inside*. Sawara Cypress (p.28) and Taiwan Cypress (opposite): sharp-tipped side-leaves but much finer, in meagre, less weeping sprays. Lawson Cypresses (p.25): large leaves of vigorous young trees faintly silver-marked beneath, and never so long or sharp.

VARIANTS Afghan Hound Tree, 'Pendula' (1884), carries *curtains* of foliage from few, *gaunt* branches; very occasional.

OTHER TREES Kashmir Cypress, *Cupressus cashmeriana*, is the most desired true cypress with comparable 2D sprays of long-pointed leaves (cones 18 mm, ripening in two years) – tender and very rare outdoors. The *clear turquoise foliage* is *spectacularly weeping*, though odd trees surviving in colder regions can be a sorry sight.

Leylandii ✪ 🌢

× Cupressocyparis leylandii

(*Cupressus × leylandii*) Nootka Cypress has hybridized several times in Britain or Ireland to make trees of huge hybrid vigour. The Leylandii – at once the *most planted* and most hated garden tree – is a cross with Monterey Cypress which seems to have first occurred, unnoticed, at Rostrevor (Co. Down) around 1870. The clone 'Rostrevor' remains rare; the easily propagated, ubiquitous 'Haggerston Grey' arose at Leighton Park, Powys, in 1888, when a Monterey pollinated a Nootka Cypress. Very occasional in forestry. Birds, at least, love it.

LEYLANDII
'HAGGERSTON GREY'

APPEARANCE Shape Very *densely columnar*, generally clothed to the base and tapering to an open, slanted tip; dark sooty-green surface, with little texture. Grows rapidly to 30 m in any soil, then blows down. **Bark** Dull red-grey; vertical/criss-crossing shallow stringy ridges. **Leaves** In rather '3D' plumes – often flattened in the ultimate two divisions, which are *at right-angles* to the third division; the 3 mm scales have sharp straight tips and no white markings underneath. **Flowers** Often *absent*, though 'Leighton Green' (below) can have massed yellow male flowers in summer and brown 2 cm cones.

COMPARE Mexican Cypress (p.33): finer, fluffy foliage.

VARIANTS 'Leighton Green' occurred – also, strangely, at Leighton Park – in 1911 when a Nootka pollinated a Monterey Cypress. Sprays *flatter* (cf. Nootka Cypress; broader; dense columnar crown distinguishes). A brighter-green tree but harder to propagate, so now rather rare.

'Naylor's Blue', a sister-tree to 'Leighton Green', was propagated only when a whirlwind demolished it in 1954; rare. *Dark softly blue-grey* foliage, in rather 3D plumes; more raggedly open-crowned.

'Stapehill' ('Stapehill 20'; Ferndown, Dorset, 1940) sheds its brown inner foliage, so the *crown is open and wispy* and the rather 3D sprays appear moth-eaten: often looks moribund, but disappoints. Occasional, because it was claimed to do well on chalk.

'Castlewellan Gold' arose at Castlewellan (Co. Down) in 1962 when a Nootka pollinated a Golden Monterey Cypress. As the first yellowish Leylandii its ubiquity was assured. *Squat-conic*; 3D plumes gold-tipped in spring/summer then *dull olive*. 'Golconda' (1977) is *bright yellow*; now in many hedges.

'Harlequin' (1975; occasional to date) has many 1–10 cm pale yellow patches; 'Silver Dust' (1960) has rather fewer 1–5 cm white shoots.

'True Cypresses' (20 species) have larger cones, which generally take two years to mature, than those in the *Chamaecyparis* genus. The foliage of the commoner species is held in 3D sprays, and the leaves are often blunt. (Family: Cupressaceae.)

Monterey Cypress 

Cupressus macrocarpa

Confined in California to two windswept cliff-tops, but since 1838 much the commonest and biggest true cypress near S/SW coasts of Britain and Ireland: it is very salt-tolerant. Less frequent inland. Sometimes seeding. Seldom in plantations and now rare in hedges.

APPEARANCE Shape Typically barrel-like, on many steep limbs from a *short bole*; near the sea often very broad, or shaped like Cedar of Lebanon; often dying back (*Corinium* canker). *To 40 m: vigorous, wispy, sharp plumes* from the crown help distinguish it from other cypresses/junipers. **Bark** Grey-brown; stringy criss-cross ridges. **Leaves** In dark green 3D sprays, with appressed scales whose tiny points can hardly be snagged by running a nail down the shoot; no white markings. Rich, lemony aroma. **Cones** Shiny, 3 cm, the scales not sharply ridged; *seldom conspicuous* except on dying trees.

COMPARE Italian Cypress (p.34) and Rough Arizona Cypress; West Himalayan Cypress, Gowen Cypress; Pencil Cedar. All have smooth, 3D shoots but are slower/narrower trees.

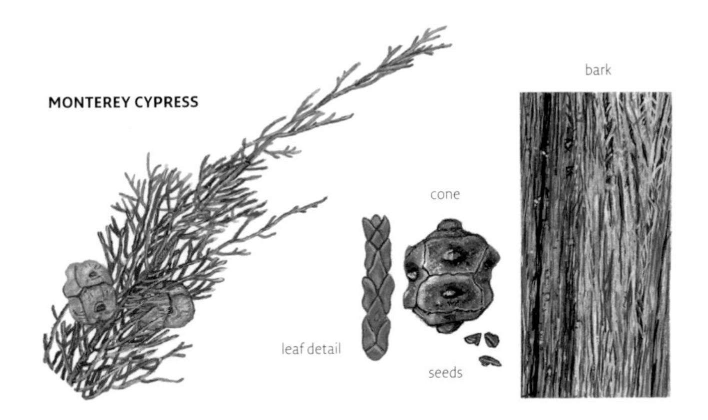

VARIANTS 'Lutea' (1892) has laxly spiky *lime-green* plumes and is frequent as an old tree, especially on coasts where it copes especially well.

'Goldcrest' (1946) is the most abundant of the newer, *acid yellow* clones. *Narrowly columnar* when young, often *slanted*; broader with age, with sharp dense turrets towards tip, often *twisted as if wind-blown*. Young pot-plants carry many soft 5 mm juvenile needles.

'Donard Gold' (1935; rare?) is slightly duller and broader, with more spreading turrets.

'Coneybeari' makes a giant broad bush with dull yellow *thread-like hanging sprays*. Rare (cf. Sawara Cypress 'Filifera Aurea', p.29).

'Fastigiata' is *very narrowly* and neatly but sinuously columnar with an open, pointed tip; rare (cf. Italian Cypress, p.34).

Mexican Cypress

Cupressus lindleyi

(Often equated with the Cedar of Goa, *C. lusitanica*, a clone – of Indian origin? – cultivated in Portugal by 1634.) Central and S Mexico; Guatemala. Occasional in milder areas.
APPEARANCE Shape Rather *densely columnar* until old, but a *fluffy*, ragged outline, on a generally *pole-like* stem with stringy coppery-grey bark; to 30 m. **Leaves** *Sharp, spreading (but very fine) tips* (unlike other true cypresses' with 3D sprays). Variably grey (though without white markings); almost scentless. **Cones** Only 15 mm wide; a prickle on each scale.
COMPARE Leylandii (p.31): harder, closely set, dull dark sprays, dull bark; seldom a good stem.

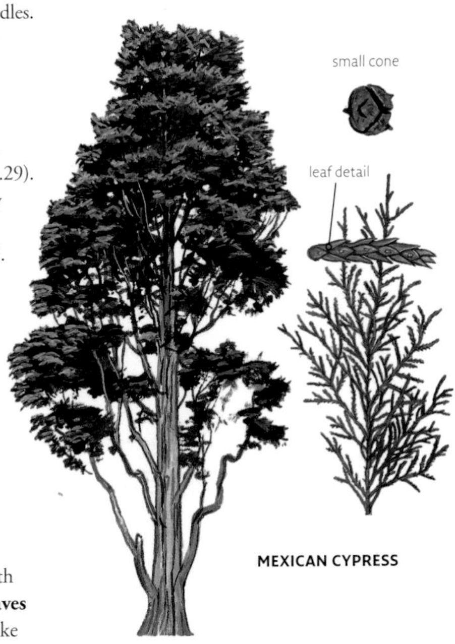

VARIANTS 'Glauca Pendula' has blue-grey, *tumbling plumes*; columnar or sprawling. Spectacular; perhaps now the most commonly planted form.
OTHER TREES Bentham's Cypress, *C. benthamii* (NE Mexico, *c.*1838; often treated as a variety) has bright, shining green foliage in *flatter, fern-like* sprays from *sinuous* branchlets; the leaf-tips *scarcely spread*. Confined to collections in mild areas. (Fluffier-crowned on a stronger stem than false cypresses, or West Himalayan Cypress.)

Italian Cypress

Cupressus sempervirens

E Mediterranean to Iran, but long grown further W and N: the definitive Mediterranean landscape tree. Rather occasional (but hardy to NE Scotland): small gardens, parks and churchyards in warmer parts.
APPEARANCE Shape The variably densely columnar f. *stricta*, with steep branching and no persistent trunk, is the familiar form, to 24 m. Var. *horizontalis* is often unrecognized: spreading branches and a lumpy, irregular crown as broad as high. Can be blackish or greyish green. **Bark** Grey-brown stringy ridges, like many true cypresses. **Leaves** Dense, 3D sprays of dark leaves with closely appressed tips and no white markings. Sweet, resinous scent may be very faint. **Cones** 3 cm, dull *grey*, each scale with a *blunt knob*; abundantly *studding the dark crown* for a year.
COMPARE Monterey Cypress (p.32). The rare Monterey Cypress 'Fastigiata' and the spreading form of Italian Cypress make identification tricky. The more numerous, grey cones of Italian

variable habit

'GREEN PENCIL'

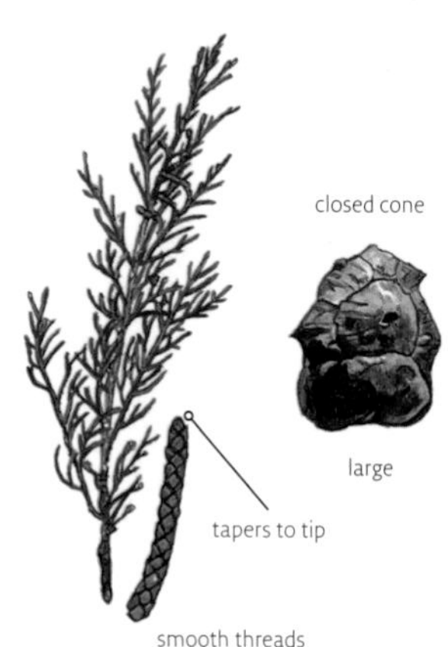

Cypress, with knobbly scales, are helpful, as are the more vigorous, plumy growths of Monterey Cypress. Monterey Cypress shoots are faintly *club-ended*; in Italian Cypress they *taper*. Chinese Juniper (p.37) has thicker foliage threads.

VARIANTS 'Green Pencil' is an absurdly tight recent selection of f. *stricta*. Only the as yet very rare Tibetan Cypress, *C. gigantea*, is as narrow.

Smooth Arizona Cypress

Cupressus arizonica var. *glabra*

(*C. glabra*) Central Arizona. 1907. Now very frequent in parks and small gardens in several selections.

APPEARANCE Shape Conic, to 22 m, with a narrow tip or egg-shaped in age; edges somewhat spiky with dense, rugged, round-ended plumes. **Bark** Beautiful: purples and reds, flaking in smooth circular or snakeskin scales. Old trees often conspicuously grafted on the type. **Leaves** In bright pale grey 3D sprays, the appressed leaves *often with a central white spot of dried resin.* **Male flowers** Abundant and yellow all winter. **Cones** 2 cm, retained for many years.

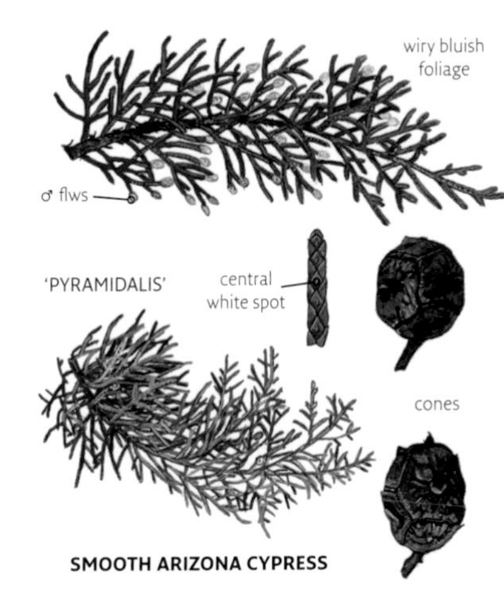

SMOOTH ARIZONA CYPRESS

VARIANTS 'Pyramidalis' (1928) is probably the common clone today: tidily shaped, with white spots on *nearly every leaf*; 'Hodgins' can be narrower.

'Aurea' has *yellow* young foliage slowly turning pale grey – rather rare but striking.

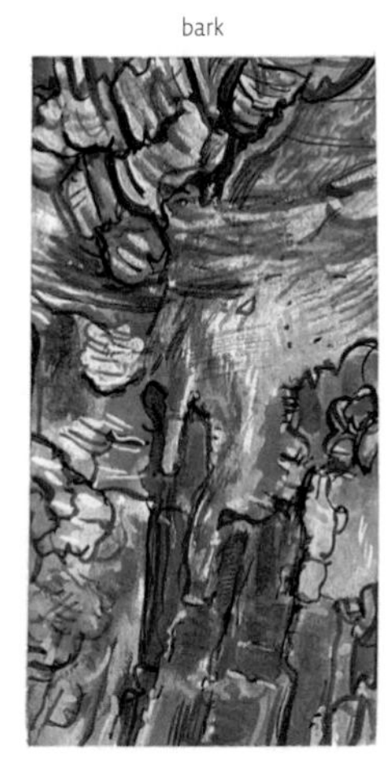

JUNIPERS

Many junipers (50 species) are bushy. They are the only 'cypresses' with berries (formed by the fusion of fleshy cone-scales). Most retain juvenile leaves – straight, sharp, often in threes, in 3D sprays; their narrow bases do not hide the quickly brown shoots. (Family: Cupressaceae.)

Common Juniper ✪ ⬮

Juniperus communis

The only conifer native both to Eurasia (including Britain and Ireland) and to North America, N to the Arctic Circle; also in N Africa. In Britain, a scarce plant of chalk scarps, open moorland and Scottish pine woods, dwindling through changes in land management (it needs a sudden cessation in grazing to colonize grassland, before taller competition shades it out or makes it vulnerable to fungal infections). Rarely planted.

APPEARANCE Shape A softly greyish bush, to 8 m. Dense and varied, like a topiary menagerie. **Bark** Grey-brown; very shaggy and stringy. **Leaves** All juvenile: to 1 cm, with *one white band* on the cupped inner side; in threes up shoots red-brown after one year. Scent of apples/lemons. **Flowers** Dioecious. **Fruit** Berries, 7 mm, ripen black in three years, with 1–3 seeds: the flavouring for gin.

COMPARE Syrian Juniper (below): much longer leaves. Meyer's Juniper (p.37): weeping habit. Taiwania: needles whose broad bases *clasp* its shoots. Sawara Cypress 'Squarrosa' (p.29): softer leaves (in *pairs*).

VARIANTS Irish Juniper, 'Stricta' ('Hibernica'; 1838), is an occasional, tight, grey selection

fruit takes 3 years to ripen

(to 8 m), with upright branching *and shoots*; Swedish Juniper (f. *suecica*) differs in *nodding* shoot-tips.

'Oblonga Pendula' (a few collections) has hanging shoots and leaves twice as long and *scarcely white* on the inner side.

OTHER TREES Prickly Juniper, *J. oxycedrus* (mountains and dunes of S Europe to Iran; in some N European collections), has 10–25 mm leaves, often with *two grey bands* on the inner side. Its berries, *6–15 mm and usually brownish*, ripen in two years.

Meyer's Juniper

Juniperus squamata 'Meyeri'

An old Chinese selection of an undistinguished Asiatic bush. 1914. Abundant; to 8 m.

MEYER'S JUNIPER

APPEARANCE Shape *Dark blue*, arching, gaunt sprays *like pampas grass plumes.* **Bark** Very shaggy; brown. **Leaves** Dense and juvenile: 8 mm, spiny; two brilliant blue-white bands on the inner side.

MEYER'S JUNIPER

Chinese Juniper

Juniperus chinensis

China, Japan. 1804. Occasional as an older tree. Schizophrenic: *tufts of juvenile foliage* can persist often at the *base* of adult, scale-leaved, 3D sprays.

APPEARANCE **Shape** A tight column, or egg-shaped, or rugged; to 20 m, on fluted/fused stems. Dingy out of flower.

CHINESE JUNIPER

Bark Grey-brown; stringy, twisting strips. **Juvenile leaves** To 1 cm, sharp, rigid, grey-blue inside; densely in twos/threes. **Adult leaves** Cypress-like, their *pale blunt margins* making a faint pattern; in *quite thick* (1.8 mm), smooth threads; catty smell. **Flowers** Dioecious: male trees yellowish through winter with flower. **Fruit** Females carry *6–8 mm* berries (whose separate scales remain visible), ripening from whitish to dark brown in *two years.*

COMPARE Without juvenile foliage/berries, a lumpier tree than Monterey Cypress (p.32) and Italian Cypress (p.34), *never with woody cones* or open plumes of vigorous growth.
VARIANTS Golden Chinese Juniper, 'Aurea' (1855), is occasional, to 14 m: bright and usually neatly conic. Male (so no berries) but much less spiky than Monterey Cypress 'Goldcrest' (p.33).

'Kaizuka' lacks juvenile foliage; a female plant, berrying freely. Nightmarish and broad: long, *gaunt, wandering, twisted turrets*. Rare.

'Keteleerii' makes a *dense greyish spire*, again with only adult foliage but plenty of berries. Very rare.

The redwoods include 17 conifers, often gigantic and with spongy red bark. They survive in scattered montane populations in both hemispheres. (Family: Cupressaceae.)

Summit Cedar

Athrotaxis laxifolia

The least rare (in big gardens) of three species from the mountains of W Tasmania. It is intermediate, but shows no obvious signs of hybridization.

APPEARANCE Shape Sturdily open-conic, to 20 m. **Bark** Red-brown: hard, shaggy ridges. **Cones** 2 cm; abundant. **Leaves** Similar to those of Japanese Red Cedar (p.39), but much shorter (4 mm free), and more broadly tapered, flatter and harder, with *hooked* tips; green on *both sides* (some small, whitish zones). Young growths are *yellow*.
COMPARE Patagonian Cypress: blunter leaves strongly white-marked on each side. Giant Sequoia (p.41).

SUMMIT CEDAR

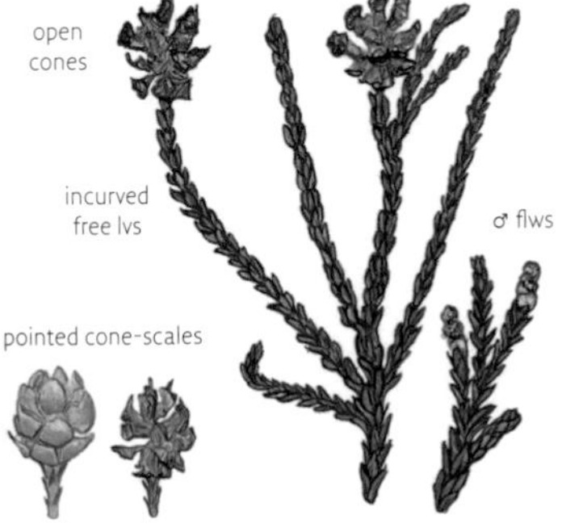

Japanese Red Cedar
Cryptomeria japonica

A giant tree in the mountains of Japan. Rather occasional; rare in dry/cold areas; in a few forestry plantations.
APPEARANCE Shape Conic on a straight stem to 40 m (occasionally layered or low-forking); rather *patchy*, rich, hard green foliage. **Bark** Orange-brown; the soft, stringy ridges much flatter and wider than Giant Sequoia's (p.41). Trunk cylindrical but sometimes carrying burrs like great beer-bellies. **Leaves** *Slightly curving*, 10–15 mm, one by one all around the drooping shoots, which their broad bases cover; four-sided, the inner two with grey-green bands. **Cones** 2 cm – abundant on some trees, absent on others.
VARIANTS Chinese Red Cedar (var. *sinensis*; S China) is sometimes considered a separate species (Fortune Cedar, *C. fortunei*) and reached Britain first (1842). Leaves slightly longer, in *soft, dishevelled, slightly weeping,* paler green sprays; scarcer as a younger tree than the Japanese form. Pre-1878 trees are either var. *sinensis* or 'Lobbii', which conveniently represent extremes of habit.

'Lobbii' is a clonal name sometimes used for trees from Thomas Lobb's first consignment of Japanese seed (via Java) in 1853: particularly short leaves in dense *tufts*, bunched on the stiff, narrow crown and with one tuft forming the apex.

'Elegans' (1861) is an abundant bush, broad and sprawling but sometimes growing as a single-boled tree, to 20 m. Cloudy dark blue-grey foliage; *purplish* – even shocking maroon – through winter, and making a fine contrast with the peculiarly bright, almost glistening orange bark. Leaves juvenile, held individually and distantly: 2 cm, *soft* and

JAPANESE RED CEDAR

CHINESE RED CEDAR

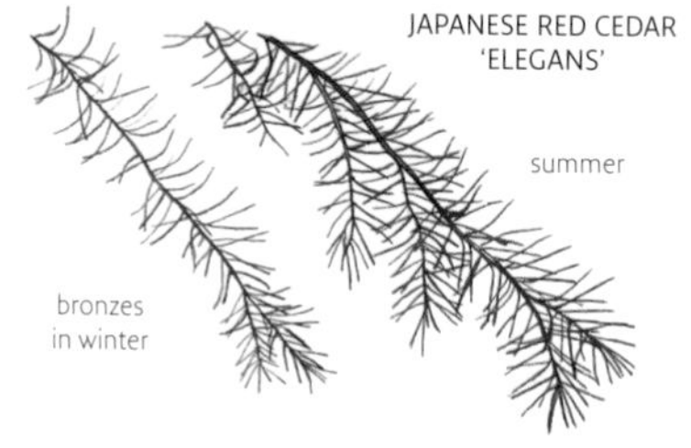

slender, rounded and curling randomly. Habit and long needles suggest a juniper. Sawara Cypress 'Squarrosa' (p.29) is an erect tree with much shorter needles in *opposite pairs*.

Coast Redwood

Sequoia sempervirens

(*Taxodium sempervirens*) The world's tallest tree now that the best Douglas Firs and eucalypts have all been felled for their superior timber. Confined to a fog-bound coastal strip from just into Oregon to S of Monterey, California. 1843. Rather occasional, and rare in dry/exposed areas; in one or two forestry plantations. Unlike most conifers it coppices happily and the rare fallen tree can grow a line of new ones; it very occasionally self-seeds in Britain.

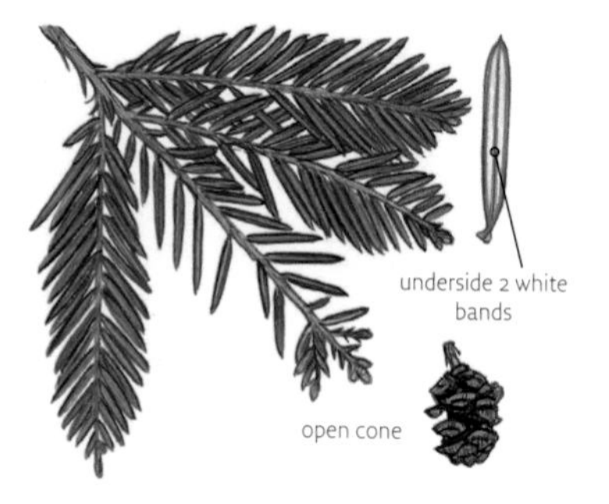

APPEARANCE Shape A ragged-sided, dark, misty-leaved column, often broad: sparsely spire-tipped while adding height, then flat-topped; to 45 m only in deep shelter. Turns brown in dry winters, without permanent injury. **Bark** Rufous, deeply spongy; hard purplish surfaces develop in age if few squirrels climb the trunk. **Leaves** Young stems *radially clothed with big scale-leaves,* which have free sharp tips. Side-shoots with yew-like needles in flattened ranks: hard, straight; to 2 cm but *progressively shorter* at shoot tips and bases; two white bands beneath. **Cones** 2 cm: unimpressive.
COMPARE Common Yew (p.17). Allies of yew have needles bigger or with no white bands underneath – and no scale-leaves up the shoots.

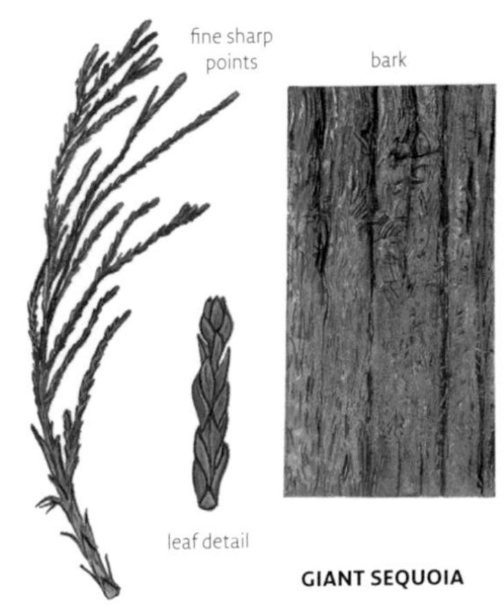

VARIANTS 'Adpressa' ('Albo-spica'; 1867) has small short leaves (cf. the Common Yew 'Adpressa', p.18), white-banded above and *entirely cream-grey on new shoots*. Rather rare: a thin, often gaunt tree, to 22 m.

'Cantab' has much *broader, very blue-grey* leaves. First distributed (1951) as a prostrate dwarf, but readily sending up vertical shoots which grow as vigorously as the type's; to 20 m so far. Rare.

Giant Sequoia

Sequoiadendron giganteum

(Wellingtonia; Big Tree; Giant/Sierra/Californian Redwood; *Sequoia gigantea*) Another Californian 'living fossil' (from a few groves high in the Sierra Nevada); 'General Sherman' is believed to be the world's largest tree, though a Coast Redwood of similar proportions was 'discovered' in 2000. With their similar names the two trees are much confused, but differ widely. 1853. Already the largest tree in all parts (to 52 m); quite frequent (away from the driest or most polluted lowlands) and very conspicuous, though highly sensitive to salt spray. Never in plantations: useless soft timber, like cardboard.

APPEARANCE Shape A dense, dark blue-green spire while adding height, with light branches, *swooping* and upcurving; the oldest trees are now subject to middle-aged

spread and have rugged, broken, columnar crowns, except in deep shelter. May shatter in high winds but very seldom blows down. **Bark** Dark red, thickly spongy (proofing the tree against forest fires); harder and darker with age. Bole nearly always *very flared.* **Leaves** Rather inconsequential – sharp radial scales on cord-like shoots, 4 mm wide, with an aniseed scent and no white markings; their tips may curve *outwards,* unlike those of Summit Cedar.
Cones Disappointing: 4 cm.
COMPARE Summit Cedar (p. 38).

Swamp Cypress
Taxodium distichum

(Bald Cypress) Texas to New Jersey. Rather occasional in warm areas. Can cope with waterlogged conditions, throwing up 'knees' to trap silt around its roots (and probably to help the roots to breathe), and it is often assumed to need water. In fact it grows faster in a well-drained soil and thrives on dry sands. Long-lived and healthy, hardly ever blowing down, though trunks and branches easily shatter: one or two English trees are probably 1640 'originals'.

APPEARANCE Shape The European population is now very diverse: sometimes a dense, solid, rounded spire on a sinuous bole; sometimes broad with heavy limbs, and often leaning; sometimes open and irregular with lax, fluffy horizontal sprays; sometimes with distant, weeping branches. To 35 m. A deciduous conifer, bare and twiggy until late spring, then fresh green, and dark red-brown late in autumn – the tip still green at Christmas. **Bark** Pale orange-grey, shallowly stringy. **Leaves** Main shoots and tip-growths have spirally set leaves with short free apices. On the short side-shoots – which are borne alternately along the twigs and drop whole in autumn – the leaves are to 2 cm long, in two flat ranks. They are grey-banded beneath, and much softer and fresher green than Coast Redwood's (p.40). **Flowers** Often dioecious. Male catkins, to 20 cm, prominent through winter after hot summers then shedding pollen in mid-spring. **Cones** 3 cm, ripening in one year.

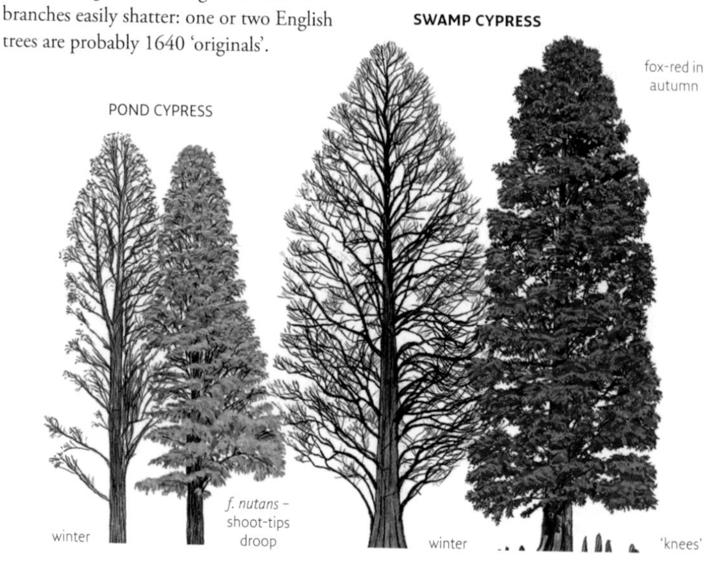

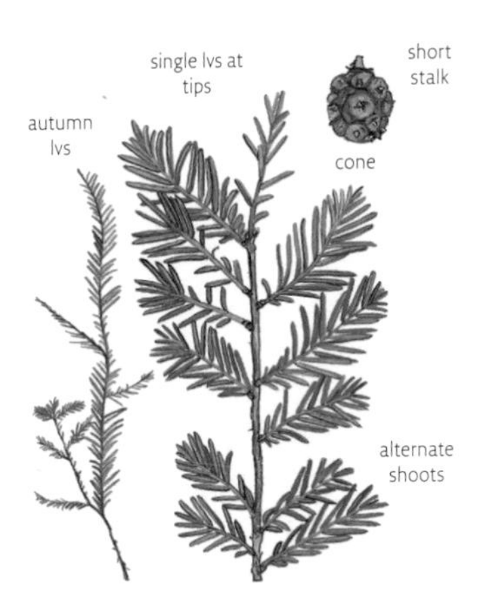

COMPARE Dawn Redwood (below). In winter, softer, less craggy bark than larches' (pp.55–7) and less spiky twiggery.
VARIANTS Pond Cypress, var. *imbricatum* (*T. ascendens*; Virginia to Louisiana) is a smaller, less hardy tree, to 22 m. Rather rare and, in summer, quite distinct: the side-shoots *rise vertically* from the branches in a pale green mane, *all closely clothed with 8 mm spirally set leaves*. The shoot-tips *nod* by late summer in the narrowly columnar selection f. *nutans*. Cannot in winter be told from poor, slender examples of the type.

Dawn Redwood

Metasequoia glyptostroboides

(Water Fir) SW China; a critically endangered giant tree in the wild, discovered only in 1941. Other conifer genera have been introduced at least as recently as this – in fact they continue to be discovered – but the Dawn Redwood is unique in its instant European celebrity, its ease of propagation from cuttings, and the success with which it has grown in all warmer regions: now locally frequent except in the smallest gardens.
APPEARANCE **Shape** A consistently dense spire so far on a *straight trunk* (forked trees are very rare), rounding off in exposure; to 30 m. In dry open sites the lower trunk often becomes extraordinarily flared and *convoluted*. Earlier in leaf than Swamp Cypress; similar (but earlier) autumn colours. **Bark** *Darker* red than Swamp Cypress's and spongier. **Leaves** Comparable in design to Swamp Cypress's, but the side-shoots (and winter buds – carried beneath the shoots – and the individual leaves) come in *opposite pairs*. Leaves longer (3 cm) and broader, and a heavier richer green; grey-green underneath. **Flowers** Seldom seen in N Europe.
VARIANTS 'Gold Rush' (2000, from Japan) has *soft, yellow foliage*.

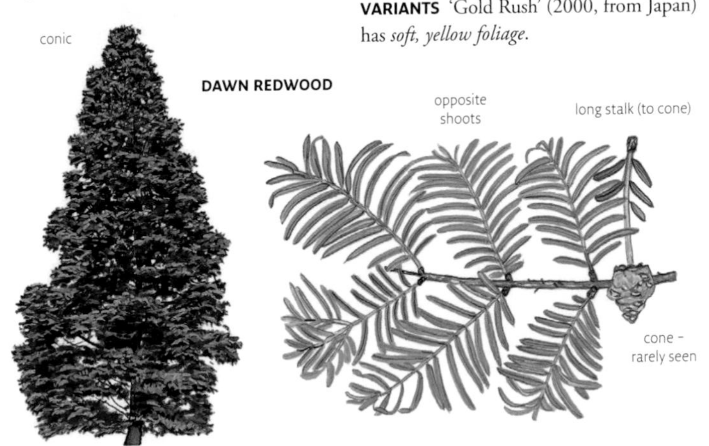

SILVER FIRS

Pinaceae is a family of conifers with needles, on twigs that soon become woody and brownish. The needles of silver firs are attached to the smooth shoots by little green sucker-like bases; most will grow in Northern Europe, but tend to be intolerant of dry air or pollution.

Things to Look for: Silver Firs

- Shape: How narrow-topped is the mature tree?
- Bark: How rugged/flaky? Is it grey, black, orange? Are there rings round the branch-scars?
- Trunk: Is it fluted?
- Shoots: How hairy and what colour? Orange, white, greenish or purple?
- Buds: How long and how resinous? What colour?
- Leaves: Distribution around shoots? How perpendicular to them? Are they curving, notched, rounded, spiny? What white bands/marks (especially above; how bright?)?
- Cones: Do bracts project above each scale?

Key Species ✪

(Leaves tend to be more flat-ranked in shade, and always rise above flowering shoots at the tree-top.)
Grand Fir (p.46): leaves in *flattish ranks* either side of the (non-flowering) shoots. **Noble Fir** (p.49): leaves *sweeping out and up* above the shoot, like a Mohican haircut. **Caucasian Fir** (pp. 45–6): straight leaves, *covering the shoot above but with very few below.* **Grecian Fir** (p.47): almost perpendicular leaves *all round the shoot.*

European Silver Fir
Abies alba

(Common Silver Fir) High ground in France and Corsica; the Pyrenees; the Alps, S and E through the Tatra to the Balkan mountains. Brought to Britain in 1603; becoming rare here except in upland regions where long-lived and still frequent. No longer used in forestry, as in the mild, oceanic climate it has become very vulnerable to defoliation by the aphid *Adelges nordmannianae*. Sometimes naturalizing in humid areas.
APPEARANCE Shape In youth, a regularly whorled spire, the stout leader seldom broken. Old trees sometimes straight, more often heavy-limbed and with *broad*, slanting or fanning tops; dull, *dark* crown,

often scraggy and never really luxuriant. A towering tree, which has reached 55 m in Scotland and even more in Germany. **Bark** Shades of grey. Soon cracked into small squarish plates, but sometimes *scalier* than other firs, and suggesting a species of spruce. The bole is usually untidy with dead snags. **Shoots** Usually *dull* grey-buff, with variably dense tiny dark-brown hairs. **Buds** Red-brown, *scarcely resinous*. **Leaves** 2–3 cm; round-tipped or (like the leaves of most silver firs on non-flowering shoots) neatly notched. They are angled forwards at about 30°, in approximately flat ranks on shaded lower branches (cf. Grand Fir, p.46 – but *much shorter*); on flowering shoots they curve and spread above the shoot (like all the leaves of Caucasian Fir). **Cones** High on old trees: brown, with the bract-scales projecting for 6–7 mm and downcurved, like panting dogs' tongues. Like all silver fir cones, they disintegrate on the tree, so are of limited use in identification.

COMPARE Grand Fir (p.46): longer, very flat-ranked needles. Caucasian Fir (below): leaves always spreading above the shoot and brighter white beneath, conspicuously silvering the branch-ends as you look up; a more luxuriant, glossy tree with less 'scaly' bark).
VARIANTS 'Columnaris', derived from French Alpine populations, has *steeply rising branches*; scarcely grown in the UK.
OTHER TREES Sicilian Fir, *A. nebrodensis*, is one of the world's rarest trees: in the 1970s the wild population was down to 21 examples on Mt Scalone, where great forests had stood. Conservation measures are now allowing regeneration to take place, and it is also growing in several UK collections, to 17 m so far. Buds red-brown, with *much white resin*. The leaves (spine-tipped on vigorous growths) tend to curve upwards and spread across the top of all the shoots.

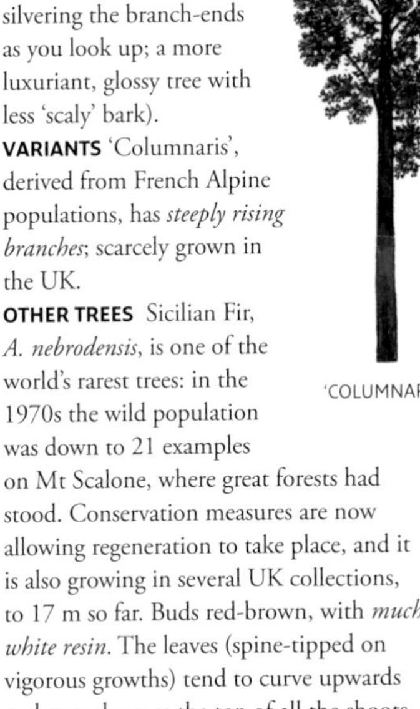
'COLUMNARIS'

Caucasian Fir ✪ 💧
Abies nordmanniana

(Nordmann Fir) NE Turkey and W Caucasus Mountains, where it is Europe's tallest tree (to 70 m). 1848. Occasional and sometimes in small town gardens: more

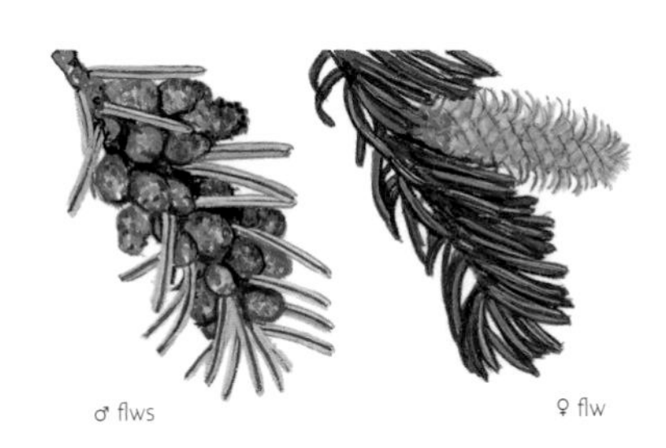

frequent to the N and W. Upmarket Christmas tree, so now in a few plots (the leaves can persist for 25 years so do not fall off in the 12 days of Christmas).

APPEARANCE **Shape** Nearly always with a *straight trunk to the tip*. Narrow, dense; columnar with age; to 50 m so far. **Bark** Mid/pale grey; shallowly and tidily square-plated with age. **Shoots** Greenish grey-brown, often with tiny dark hairs. **Buds** Pale brown, not resinous. **Leaves** Glossy mid-green; held forwards at 30°; all across the top of most shoots, though with a gap at the vertical on weak, shaded ones; few or none below. Clear white bands underneath; usually notched at the tip; light fruity/petrol smell. **Cones** Bract-scales exposed and bent back for 2 cm.

COMPARE Douglas Fir (p.66): also with this foliage arrangement, but with slender, long-pointed buds and much softer, slenderer leaves. European Silver Fir (p.44): most leaves often more or less flat-ranked; a rougher, darker, scalier tree.

CAUCASIAN FIR

Grand Fir

Abies grandis

(Giant Fir) Vancouver to California. 1832. Rare in dry areas but frequent in the N and W; occasional in plantations; sometimes seeding. The most vigorous silver fir in the UK and the tallest tree in many areas, to 62 m.

APPEARANCE **Shape** A narrow spire in youth; often rising high above other trees and much broken or lop-sided with age, but with vigorous new tips regrowing close together, even in dry districts, or a single slender wandering top. **Bark** Silvery- or purple-grey, cracking with age into quite close rectangles. **Shoots** Brownish *olive-green* (dull greyish rufous by the second year). **Buds** Like *tiny* (2 mm) grey pearls. **Leaves** Long (25–50 mm) and more strictly *flat-ranked* than any other fir's (but curving above the flowering shoots). Narrow white bands beneath. Delicious tangerine-rind scent. **Cones** Small (8 cm) and high up, on old trees; bracts hidden.

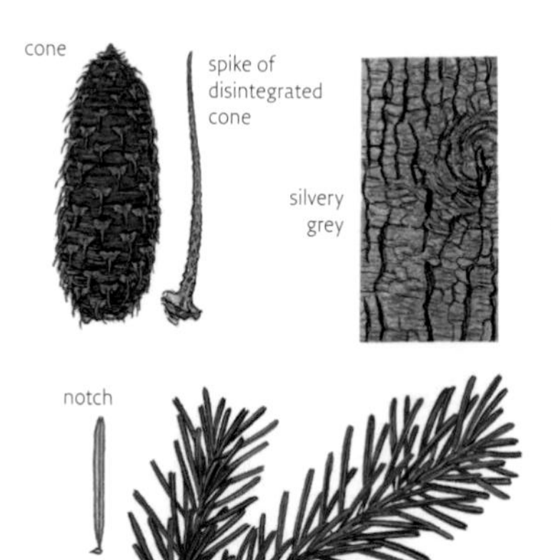

CAUCASIAN FIR

GRAND FIR

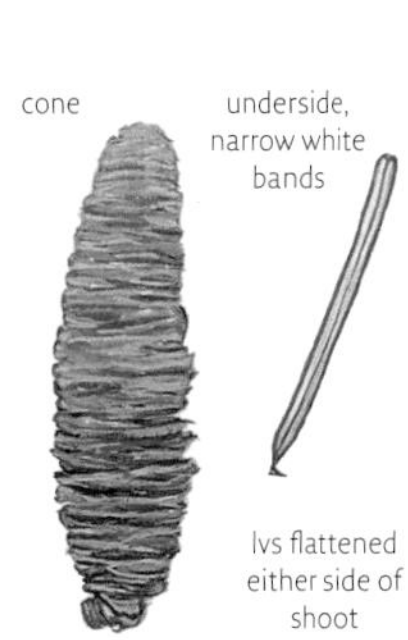

GRAND FIR

APPEARANCE Shape Seldom spire-like for long: *massive low branches* and a lumpy, often crooked trunk make a broad, much-broken, rather spiky crown, to 40 m. **Bark** Dull grey; soon closely and deeply cracked into particularly knobbly squares. **Shoots** Stout, hairless, shiny pale brown; often little seen between

COMPARE European Silver Fir (p.44): also has approximately flat-ranked but much shorter leaves; a darker, scalier tree with grey-brown shoots, larger brown buds, and projecting bract-scales. Greenish shoots and ranked leaves recall nutmegs; needles never notched).

Grecian Fir

Abies cephalonica

Greek mountains. 1824. Very occasional: coping well in dry areas though troubled by late frosts; rarely seeding.

the leaves' big 'suckers'. **Leaves** *Radiating stiffly all round the shoot* (slightly more above than below); 2–3 cm. Shiny green above with two *narrow* whitish bands beneath, leathery, dully *spined*, and balsam-scented. Their broad angle makes the sprays wider than European Silver Fir's. **Cones** Abundant high on old trees, 15 cm; bract-scales project and bend back.

COMPARE Spanish Fir (p.48): the most similar of several silver firs with leaves radiating round the shoot; leaves shorter, rounded, banded/white-dusted above. Vilmorin's Fir (p.48): hybrid of these two trees; thicker, more spaced leaves than Grecian Fir, with dull pale green bands beneath. Korean Fir (p.48): short, *blunt* leaves all round the shoots. Noble Fir (p.49): variably upcurving leaves.

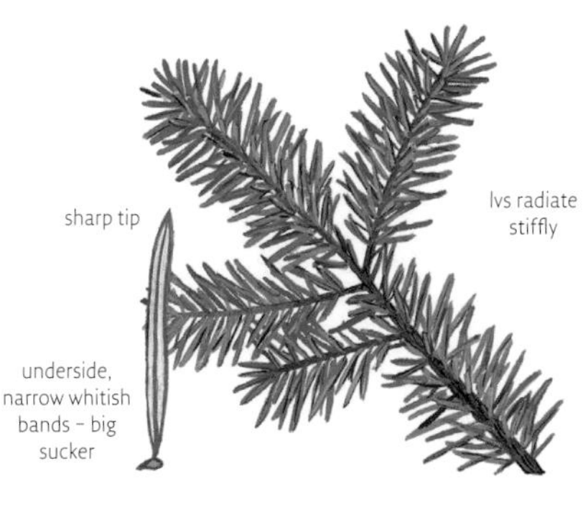

GRECIAN FIR

Spanish Fir

Abies pinsapo

(Hedgehog Fir) Confined in the wild to the shady side of a few mountain-tops above Ronda in S Spain, and very vulnerable, but one of the least rare silver firs in large gardens in the UK (since 1839), coping reasonably well with dry conditions and chalky soils.

APPEARANCE Shape Conic, to 30 m, then haphazard with age: dense and gappy, or sometimes spikily open; *greyish* – bright pale blue in grafted selections ('Glauca'). **Bark** Dark grey; developing small, craggy plates. **Shoots** Greeny-brown, the leaf-suckers often almost meeting up. **Leaves** Set densely all around the shoot (slightly parted underneath on young trees or shaded shoots); *perpendicular or leaning slightly backwards*, like an old pipe-cleaner; they are short (15 mm), thick and stiff with blunt tips and broad *grey bands on each side* (but somewhat whiter beneath in 'Glauca'). **Cones** Numerous at the tops of old trees, bright green during summer; bracts hidden.

COMPARE Grecian Fir (p.47): longer, sharper leaves, green above and narrowly white-banded beneath. Noble Fir (opposite): sometimes looks similar when foliage is out of reach; retains an often more silvery, less square-cracked bark and has sprays of longer, variably upcurved leaves (often almost solid against the light).

OTHER TREES Vilmorin's Fir (*A.* × *vilmorinii*) is an artificial hybrid with Grecian Fir (Paris, 1867), found in a few collections: a dark, *open, spiky tree*. Slightly longer (25 mm) leaves, much sparser, and sharply pointed; dark green above with pale green bands beneath.

SPANISH FIR

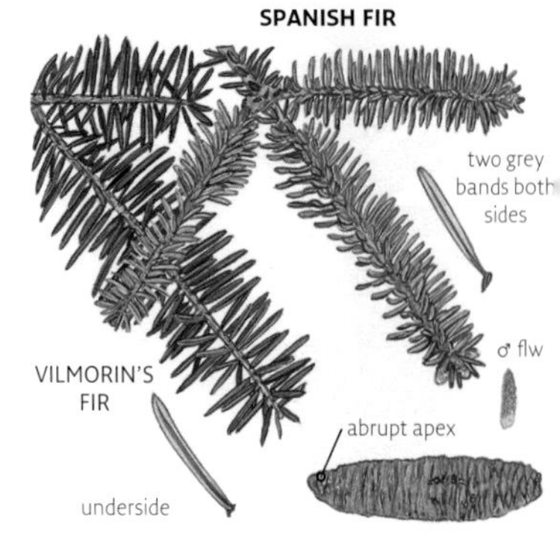

SPANISH FIR

Korean Fir

Abies koreana

Restricted to S Korea and Cheju Do island. Introduced only in 1913, but now the *most widespread silver fir*, frequent in small gardens for its display of cones.

APPEARANCE Shape Plants sourced from Cheju Do are dumpy pagoda-shaped *bushes* with crisp, closely rising branches, while trees from the mainland have reached 15 m in sheltered woodland. **Shoots** Pale pinkish grey, often lightly hairy. **Leaves** *Stumpy*

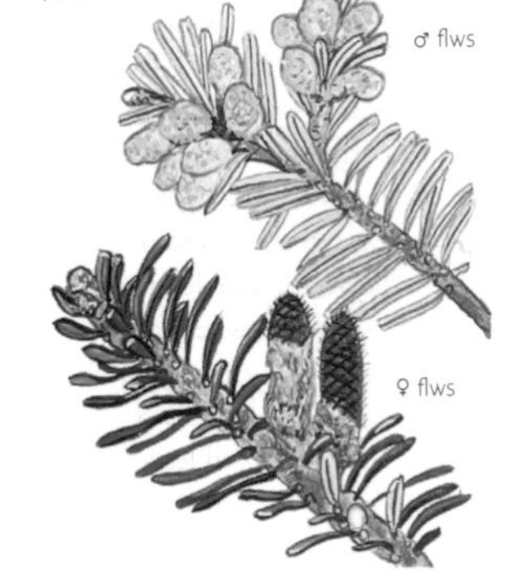

Noble Fir

Abies procera

(*A. nobilis*) NW USA. 1830. Occasional, in many large and some smaller gardens; very rarely in forestry plantations; sometimes seeding; short-lived and rare in the driest areas.

APPEARANCE **Shape** Spikily conic, to 50 m in shelter, on a bole which scarcely seems to taper. The heavy cones mean that exposed tops soon break and the tree grows *densely columnar* (sparse in shade or dry conditions). **Bark** *Silvery* or purplish grey; rather distantly cracked; some trees eventually develop rugged ridges. **Shoots** Pale rufous, hairy, but half-hidden by the *very dense leaves*, which are parted beneath the shoot or run parallel with it before

(12–18 mm), perpendicularly all round most shoots though with few below and some upper ones bent back past the vertical; usually notched. Brilliant white bands underneath *almost coalesce* and there is often a white patch above near the tip. **Cones** Abundant even on young trees; vivid purple-blue cones like Fabergé eggs. These are decorated with golden-brown, outcurving bract-scales and – like all silver firs' – ripen to brown before disintegrating on the tree, so cannot be pulled off for ornaments.

cone – purple then brown

cone-scales

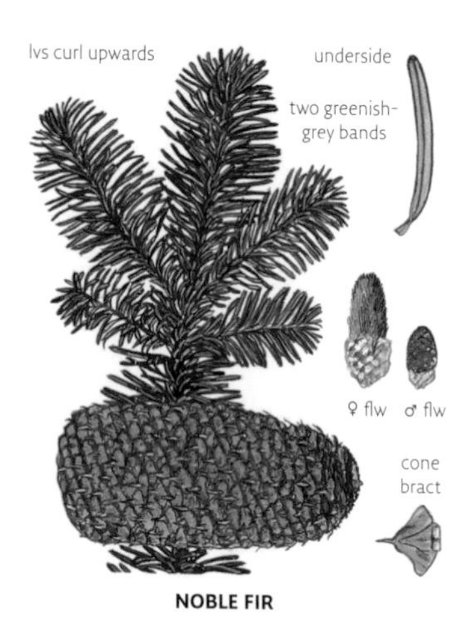

NOBLE FIR

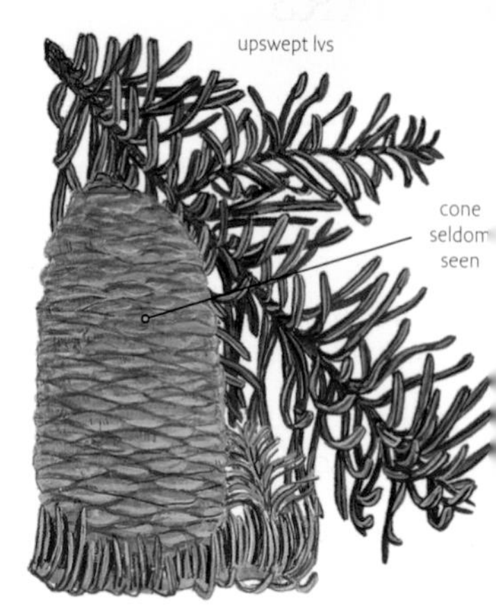

RED FIR

usually *sweeping forwards then up*; their rounded (rarely notched) tips often make a close pattern like a hedgehog's back. **Leaves** Always *grey*; grafted selections ('Glauca') are intensely silver. Slender, grooved above and flattened so that they *cannot be rolled* between finger and thumb, with two thin grey lines above and two whitish bands beneath. Rather oniony aroma. **Cones** Purple-brown with hanging, green bract-scales; ripening golden-brown; *very big, like loaves* (to 25 cm) and carried near the tops even of small trees.

COMPARE Red Fir (below): corkier bark, more formal spire shape and slenderer, almost *round* leaves. Spanish Fir (p.48): stiffly radiating leaves and square-cracking bark. Sicilian Fir (p.45): also with steeply upcurved leaves.

Red Fir

Abies magnifica

Oregon and California. 1851. (The very different *A. amabilis* is also called 'Red Fir'.) Rare: a beautiful tree, but thriving for long only in deep shelter and high humidity; the distinctively red bark of old wild trees is yet to be seen in Europe.

APPEARANCE Easily mistaken for Noble Fir. **Shape** Dense, narrow, fairy-tale spire, maintained to 45 m in Scotland, on a rather corkily ridged bole sectioned by *rings of black scars* from the *regularly whorled* branches. **Leaves** Longer, slenderer, sparser, laxer than Noble Fir's, and *round enough to be rolled*. **Cones** Bracts normally *hidden*.

corky cracked bark

bole may swell like a barrel

CEDARS

Cedars (four old-world species) carry their leaves spirally around strong shoots but in rosettes of 10–60 on very short, thick side-shoots (spurs), like the deciduous larches. Pollen is shed late in autumn. ('Cedar' is a name used for many unrelated trees with valuable, aromatic timber – including a broadleaved relative of Mahogany, Chinese Cedar) (Family: Pinaceae.)

THINGS TO LOOK FOR: CEDARS

- Shape (especially branch-tips)?
- Leaves: How long? Are they translucently spined?
- Cones: What is the shape at the tip?

Cedar of Lebanon 

Cedrus libani

The Lebanon (where much reduced by felling – the fine timber is insect-repellent), N to the Taurus Mountains. The essential accessory for a mansion lawn since 1740, when the early British trees were mostly killed by frost. Today's population is hardier. Still quite frequent in lowland areas as an old tree, though currently seldom planted. It has suffered more than most species from the misconception that it will take so long to mature that planting one is futile. In fact, it is very vigorous.

APPEARANCE **Shape** The great *level plates of foliage* develop with age in open situations. Young trees are rather gauntly conic and the rare woodland specimens have a long straight trunk, to 42 m, but a flat top. To identify a cedar it can be useful to remember 'Lebanon – level; Atlas – ascending; Deodar – drooping', but confusion is possible: Deodar *can* have big level plates, the tip-growths of Cedar of Lebanon *can* droop, and its foliage-plates *can* ascend jaggedly. A dark green tree, or sometimes very grey and as bright as many Blue Atlas Cedars (p.53). **Bark** Black-*brown*, closely ridged and cracked. **Shoots** With very fine down (less than Atlas Cedar's (p.52) and often confined to shallow grooves). **Leaves** About 25 mm, stiff; the *short point green except for its translucent extreme tip.* **Cones** Barrel-shaped, *without a sunken top*, tapering more (above halfway up) than Atlas Cedar's.

COMPARE Other cedars. Atlas Cedar (p.52): greyer bark, no very broad flat foliage-plates,

CEDAR OF LEBANON

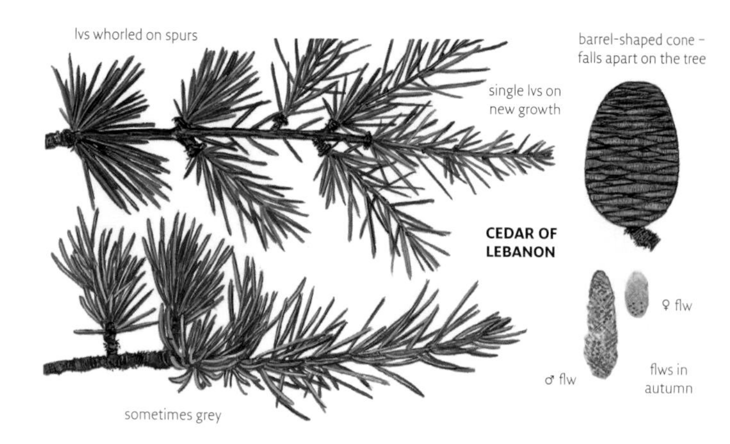

shorter leaves (to 20 mm in rosettes) with translucent spine-tips, and cones with sunken tops – but some old trees are hard to specify. Deodar (p.54): longer (30–35 mm), softer leaves, and more drooping young growths.
VARIANTS 'Aurea' is a very rare, stunted form with soft yellow foliage.

'Glauca' is a particularly *rich blue* selection; rare (cf. grey examples of the type).

Atlas Cedar

Cedrus atlantica

(*Cedrus libani* var. *atlantica*) Atlas Mountains, Algeria and Morocco. 1841. Trees of the typical green form are rather occasional.
APPEARANCE **Shape** Conic in youth; in age still tapering to *narrow*, flat top; to 38 m. The *spiky*-edged, often *rising* plates of foliage are *never as large* or flat as in good Cedars of Lebanon; young shoots rise or droop slightly, though occasional trees have strikingly weeping twigs *and branches* (some may be the clone 'Pendula'; cf. 'Glauca Pendula', opposite). Blackish to vivid green, or grey (merging into f. *glauca*). **Bark** Usually greyer than Cedar of Lebanon's, cracking often neatly into close plates. **Shoots** With short, dense, blackish hairs. **Leaves** Short (*to 20 mm in rosettes*), with small *translucent* spine-tips. **Cones** Barrel-shaped, with a *dimple at the tip*.
COMPARE Cedar of Lebanon (p.51): differences emphasized.
VARIANTS 'Fastigiata' (1890) has *steeply ascending* branches and (dark grey) sprays: a very neat tree, but rare.

ATLAS CEDAR

Blue Atlas Cedar

Cedrus atlantica f. *glauca*

Selections of the greyest wild Atlas Cedars (1845 on) are now very frequent in gardens large and small (where they grow with alarming vigour).

APPEARANCE As Atlas Cedar (opposite) but *bright silver-grey* – much paler than other cedars, and almost pink-tinted when shedding pollen in October.

BLUE ATLAS CEDAR

COMPARE Blue Colorado Spruce (p.63): some selections as bright, but never with needles in rosettes.

VARIANTS 'Aurea' has foliage *emerging pale sunlight-yellow and greying through the year*: an extraordinary 'died-last-week' effect, quite distinct from the golden Cedar of Lebanon. Rare.

'Glauca Pendula' (1900) has branches and silver-grey foliage *hanging in a single curtain* from usually one hump-backed main stem. Spectacular, like a lignified spout of water, but rare and difficult to establish.

Deodar

Cedrus deodara

W Himalayas, where it reaches 80 m. 1831. Very frequent as an old park tree and as a gorgeously soft, silky young plant in small gardens.

APPEARANCE Shape Often a very straight trunk and conic crown into maturity. Infrequently, heavy low limbs and big flat plates develop, like a typical Cedar of Lebanon's, and there may be multiple trunks from the base. Generally *vivid green*; sometimes freshly yellowish (unlike other cedars); sometimes very dark or grey. The leader *droops* (cf. Western Hemlock, p.64) and shoots always spill and hang from the edges of the foliage-plates – but less noticeably on slow-growing old trees. **Bark** Like Cedar of Lebanon's (p.51); sometimes scalier and more purple. **Shoots** Quite *densely* hairy. **Leaves** *Long, rather soft*, to 50 mm on extension shoots and 35 mm on spurs, with fine grey lines and a translucent tip. **Cones** Most like Cedar of Lebanon's, but often *absent* even from older trees.

COMPARE Cedar of Lebanon (p.51): shorter, stiffer leaves and shoots never so weeping.

VARIANTS 'Aurea' has *soft gold* younger foliage; occasional as younger tree.

'Pendula' is dumpy and *very weeping*. Rare.

LARCHES

Larches (ten species) have deciduous needles, arranged radially around strong shoots at the branch-tips and in rosettes on spur-shoots on older wood (like the evergreen cedars). Their colours are fresh green in summer, and yellow in autumn. The best means of differentiating the species are the cones, which can generally be picked off the ground all year. (Family: Pinaceae.)

THINGS TO LOOK FOR: LARCHES

- Shoots: Are they hairy? What colour?
- Cones: How big? Are the bracts visible? What colour are they when ripening?
- Cone-scales: How many? Are they straight/curved and are they hairy?

European Larch ✪
Larix decidua

(Common Larch; *L. europaea*) The Alps (replaced by Norway Spruce in colder, wetter areas), with varieties in the Tatra and Sudetan mountains and plains of Poland. Long cultivated and locally abundant: older plantations, shelterbelts and parks, away from cities and the driest areas. Rarely naturalizing. The timber is hard and rot-resistant; Tatra and Sudetan forms generally make the best plantation trees.

winter

summer autumn

APPEARANCE Shape Spire-like, on a trunk straight only in the finest, sheltered trees (to 45 m); often broad and characterful in age in dry or exposed sites. The fine shoots *hang* under the branches. *Blond* in winter: more finely and spikily twiggy than Ginkgo (p.16) or Swamp Cypress (p.42). Saplings often grow wildly twisting trunks, which straighten with maturity. **Bark** Pink-brown; wide, often criss-crossing, scaly-topped ridges. **Shoots** *Amber* or pale pinkish, hairless and unbloomed. **Buds** *Low-domed knobbles.* **Leaves** Less than 1 mm wide; vivid green; two pale bands beneath. **Female**

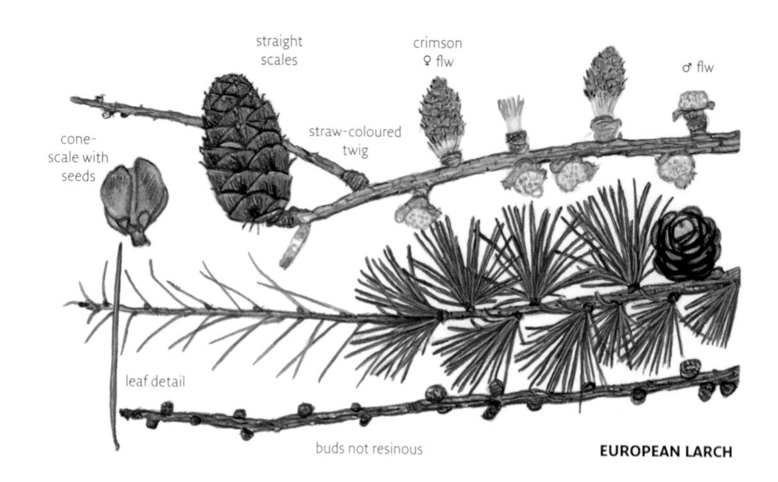

flowers As bright as rubies in mid-spring among vivid green emerging needles, but easily overlooked. **Cones** Soon brown: oval, 25–40 × 20–25 mm when ripe, the scales *not or scarcely curving.*

COMPARE Japanese and Dunkeld Larches: redder shoots; broader darker leaves; cones with outcurved scales. Western Larch and Sikkim Larch: cones with long-protruding bracts. Golden Larch: a stocky lookalike with much thicker leaves.

Japanese Larch

Larix kaempferi

(*L. leptolepis*) Mountains of central Honshu, Japan. 1861. Quite rare in gardens but a frequent plantation tree on richer soils: its particularly heavy leaf-fall is able to smother invasive rhododendron growth. In rainy climates these plantations are now being rapidly killed by *Phytophthora ramorum* (Sudden Oak Death).

APPEARANCE Shape Often broader-conic than European Larch's when open-grown, to 40 m; in winter, *smoky-red*. **Bark** Reddish or purplish grey-brown; the ridges usually become *shaggier* than European Larch's. **Shoots** *Dark* orange, brown or purplish grey, with a variable *waxy white bloom*. **Buds** Little *resinous* knobbles. **Leaves** A good 1 mm wide; *dark*, heavy green with two broad grey bands beneath. **Cones** Almost *spherical*, 30 mm, the scales *curling sharply back* like the petals of a tiny, mummified rose.

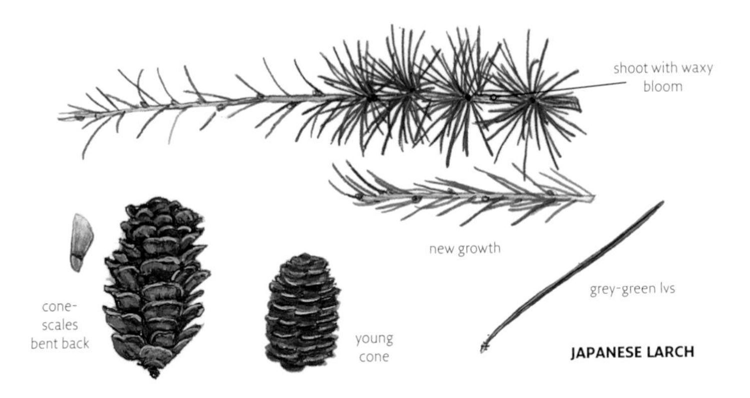

COMPARE European Larch (p.55): blond shoots, slenderer bright green leaves and cone-scales not curved back. The two trees' hybrid, below, grades into Japanese Larch. The foliage and overall appearance of most larches are very similar and exotic species in gardens are easily overlooked.

Dunkeld Larch

Larix × eurolepis

The hybrid of European and Japanese larches was first selected at Dunkeld (Perth and Kinross) in 1904, but identified later among trees in a nearby wood which had been planted seven years earlier. Rather rare as a mature garden tree (to 40 m so far), but for some decades the preferred larch in forestry.

APPEARANCE **Shape** Usually straight and often slender. **Bark** Reddish brown; most like Japanese Larch's. **Shoots** Pale orange, pale brown or reddish, scarcely bloomed. **Buds** Not resinous. **Leaves** Often longer than either parent's (50 mm); rich green above. **Cones** Oval and *tall*, to 40 × 25 mm, the scales *bending abruptly outwards to some degree but not downcurved at the rim*. Plantation trees (generally back-crosses) often tend towards Japanese Larch, with shorter cones and more curled scales, so that positive identification can be difficult.

SPRUCES

The spruces (about 40 species) are an uncomfortable bunch – thin, scaly bark, sharp needles, and woody 'pegs' protruding from the ridged shoots by which the needles are attached. The cones hang, and normally fall when ripe. (Family: Pinaceae.)

THINGS TO LOOK FOR: SPRUCES

- Shape: How broad? Does the tree weep?
- Bark: How shaggy is it?
- Shoots: Are they hairy? What colour are they?
- Leaves: Distribution around shoot? How flattened are they? Are there white bands (on which sides)?

KEY SPECIES 

Morinda Spruce (below): squarish leaves, all round the shoot and green on each side. **Norway Spruce** (opposite): squarish leaves, nearly all above the shoot and green on each side. **Blue Colorado Spruce** (p.63): squarish very grey leaves. **Sitka Spruce** (p.60): much-flattened leaves.

Morinda Spruce

Picea smithiana

(*P. morinda*) W Himalayas, on drier slopes. 1818. Very occasional: large gardens even in dry areas.

APPEARANCE **Shape** Spire-like; ruggedly columnar with age; to 38 m. *Hanging* foliage (less so than Brewer Spruce's, opposite); scraggy and moping rather than weeping in exposure. **Bark** Purplish grey; harsh round scales, or ridges like Deodar's (p.54). **Shoots** Cream, hairless. **Buds** Purple-brown, to 8 mm. **Leaves** *Longer than other spruces'* (40 mm), *slender* and sharp, *all round* the hanging shoots; squared, and dull dark green on each face. **Cones** Long (14 cm), taper-tipped; hard smooth-rimmed scales.

COMPARE Brewer Spruce (opposite): *flattened* leaves, hairy shoots. Some Norway Spruces ('Virgata'; 'Pendula') have leaves all around their strong (*orange*) shoots, but these are seldom 25 mm long. Colorado Spruce (p.63): many leaves held below most shoots but much shorter/stiffer needles.

OTHER TREES Schrenk's Spruce, *P. schrenkiana* (Kirgiziya; Tien Shan), is in a few collections, to 20 m. *Denser non-weeping habit,* bright brown buds, and shorter (35 mm), thicker, shorter-pointed leaves, fewer spreading beneath the shoot.

Brewer Spruce
Picea breweriana

Confined in the wild to a few mountain tops on the Oregon-California border. 1897. Occasional in larger gardens, and admired by many.

APPEARANCE **Shape** Stumpily conic to 20 m, with massed spiky tops, the many slender branches lugubriously hung with black foliage *like curtains of Spanish Moss*. With age often sparse or malformed. **Bark** Red-black, shedding rather large, smooth circular plates. **Shoots** Pink-brown, finely *hairy*. **Leaves** *Flattened*, 30 mm; dark, dingy green, with two narrow white bands

beneath; spread all round the shoots. **Cones** 11 cm, taper-based; slender, often curved. **COMPARE** Morinda Spruce (opposite). No flat-leaved spruce is similar.

Norway Spruce
Picea abies

(Christmas Tree; Spruce Fir) Europe from southern Scandinavia S to cold, wet mountains in the Alps and Balkans, merging in Russia into Siberian Spruce. Long grown

NORWAY SPRUCE

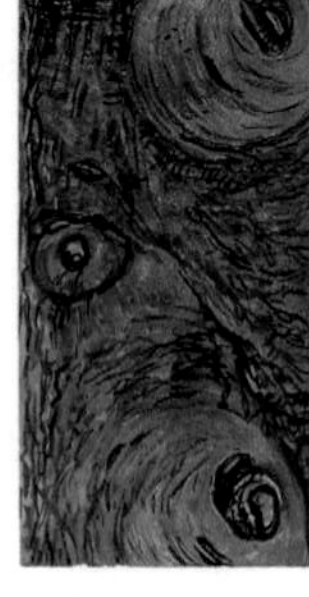
'TUBERCULATA'

in the UK; abundant in wetter areas and still much the commonest ornamental spruce; rarely naturalized. The traditional Christmas tree, regularly planted out in small town gardens though seldom thriving.
APPEARANCE **Shape** A neat spire, narrow in its dense natural stands; sometimes weeping shoots (Comb Spruce, 'Pendula'); occasionally with low limbs but never broad-topped. To 50 m in Scotland, 68 m in Germany. Heavy foliage except on sickly plants; dark rich green. Odd trees ('Tuberculata') develop hugely swollen branch-attachments (as can Sitka Spruces). **Bark** Coppery grey-brown for 50–80 years, very finely shredding; then grey/purple, cracking harshly into small rounded plates. **Shoots** Dark dull *orange*, hairless except in the east of its range. **Leaves** Quite short (15–25 mm), *hard green* with (very) faint white lines on each of the four sides (sometimes brighter beneath); stiff and pointed; spreading above the shoot but only below very strong/weeping ones. Rich, hot, sweet smell. **Cones** Long (to 20 cm), slender; scales flimsy, and jaggedly tipped.
COMPARE Colorado Spruce (p.63): thicker, spined leaves, some spreading beneath the shoot.

NORWAY SPRUCE

Sitka Spruce
Picea sitchensis

Alaska to N California, seldom far from the sea. 1831. The commonest forestry conifer in wet areas of Britain; locally frequent in parks and belts but almost absent from dry lowlands. With its massive leader, indifferent to exposure and salt spray, but stunted and yellowish when rainfall drops below 800 mm a year.
APPEARANCE **Shape** Openly conic and broad, especially in exposure. Sparse, dull blue-grey foliage: even vigorous trees can look half moribund. Huge buttressed bole when happy; to 60 m so far in Scotland. **Bark** Very purple-grey; soon with harsh, scaly plates. **Shoots** White-brown, hairless. **Leaves** Much *flattened*, with two bright blue-white bands beneath and narrower lines above; straight and ferociously spiny; to 30 mm. Very few

SITKA SPRUCE

spread beneath the shoot; *side-leaves are held perpendicular to it*; upper ones, *in marked contrast*, lie well forwards and are pressed down above it. **Cones** Short, to 10 cm; thin stiff papery scales, their margins crinkled and toothed.

COMPARE Brewer Spruce (p.59), and Serbian Spruce (below) also have flattened leaves.

Serbian Spruce

Picea omorika

Small, vulnerable population in the upper Drina Valley, Serbia/Bosnia-Herzegovina. Introduced only in 1889 but now widely planted in parks and gardens even in dry areas (where it copes well) for its generally superb shape; rarely in forestry plantations. A short-lived tree.

APPEARANCE Shape Typically a very *narrow solid pagoda* to 30 m, with shoots hanging from long-pendulous then upcurving branches, the lowest of which may layer. (The occasional broad, open specimen may not be recognized.) Rich or bluish green, the rising branch-tips brightly etched with silver from under the leaves – the sprucest of spruces. **Bark** Red-brown, developing big round scales. The foliage of the best trees needs to be parted to see the trunk. **Shoots** Pale brown, *very hairy*. **Leaves** To 22 mm; *broad* and *flattened* with (older trees) *suddenly*

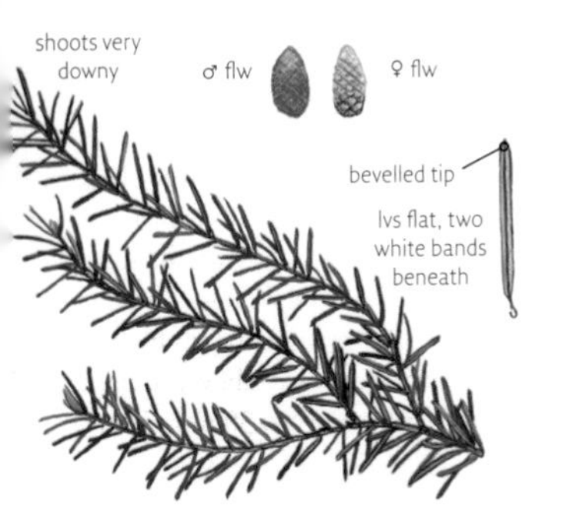

grows well in cities, often very narrow

bevelled, bluntish tips; two white bands underneath. They spread widely on either side of the shoot and are often pressed down above it. **Cones** Carried even on small trees: about 6 cm, with finely toothed scales and spots of resin.

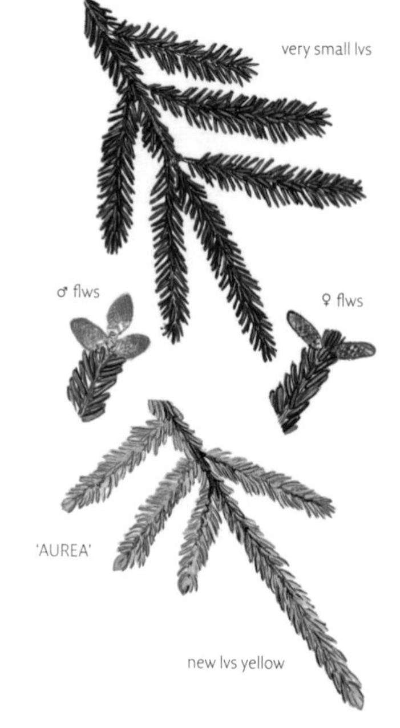

Oriental Spruce

Picea orientalis

(Caucasian Spruce) Caucasus Mountains and NE Anatolia. 1839. Very occasional everywhere, coping relatively well with dry conditions.

APPEARANCE **Shape** Neatly conic; with age, often densely columnar, the top still bristling with many fine new shoots. A dark, matt green, healthy tree, to 40 m. **Bark** Pinkish brown; very finely shredding, then developing rather regular round plates. **Shoots** Pale brown; densely *hairy* on the ridges. **Leaves** *The shortest of any spruce* (6–12 mm); square in section, *neatly round-ended* on older trees; green on each face. Very few spread below most shoots; those lying above them tend to be pressed down. **Cones** Small (7 cm); rather curved, and often spotted with resin.

COMPARE Norway Spruce (p.59): similar jizz but much longer, sharp leaves. Serbian Spruce (p.61; also with bevelled needles and doing well in towns): flattened leaves with bold white bands beneath.

OTHER TREES Maximowicz's Spruce, *P. maximowiczii* (central Japan, 1865; in a few collections) has scarcely hairy, brighter orange shoots and rather longer (10–13 mm) leaves (cf. Norway Spruce, p.59; its bark often remains a richer orange-red).

resin-encrust
cone

Blue Colorado Spruce

Picea pungens f. glauca

USA: part of a wild population with small scattered stands along the Rocky Mountains. Abundant in small gardens, dwarfed from selection or from the dry, polluted air (to 28 m in collections). The type (1862; leaves dark green/greyish) is confined to big gardens.

APPEARANCE **Shape** Columnar-conic; dense and narrow (gaunt when struggling); very spiky. **Bark** Dark purplish: coarse scales or *scaly ridges*. **Shoots** Pale brown, shiny and hairless. **Leaves** *Bright silver-blue* from whitish bands on all four sides; spreading all around most shoots but with more above them, and mostly *curving* forwards and up. To 20 mm; stiff and sharp ('pungent'). **Cones** To 12 cm; wavy, papery scales.

COLORADO SPRUCE | **BLUE COLORADO SPRUCE**

COMPARE Sitka Spruce (p.60): white bands under much-flattened leaves; grows much bigger. Blue Atlas Cedar (p.53): the only other common conifer with leaves as intensely silver as some clones; carries most needles in rosettes.

VARIANTS Silver selections include 'Moerheimii' (1912: paler shoots; leaves to 30 mm), 'Hoopsii' (*c.*1958, brightest of all; very slow), and 'Koster' (the descendants of ten trees selected in 1908).

BLUE COLORADO SPRUCE

COLORADO SPRUCE

HEMLOCKS

Hemlocks (ten species) suggest polished, non-scaly spruces (with smaller cones). Their flat, often silvery leaves have thin stalks that run parallel with the shoot. (Family: Pinaceae.)

THINGS TO LOOK FOR: HEMLOCKS

- Shape?
- Shoots: Are they hairy? What colour?
- Leaves: Notched? Minutely serrated? How broad? Any white/grey bands? Jumbled or ranked? Upside-down along the shoot-top?

Western Hemlock ✪ 💧
Tsuga heterophylla

NW North America, where it makes much the largest species. 1851. Locally abundant away from dry, polluted areas; rarely naturalizing. A frequent forestry conifer: the fine, pale lumber is also ideal for pulpwood. A tree of outstanding beauty when it thrives though plantations are singularly dark and sinister.

APPEARANCE Shape Typically straight trunk (rarely forking), fluted with age; narrow pagoda shape, to 50 m in shelter. *Hanging leader* (cf. Deodar, p.54) allows the tree to shoulder its way up through others, while the dense *downswept fans of vivid-green foliage* tolerate deep shade. **Bark** Dark brown, developing cedar-like but mor rugged ridges. **Shoots** Coffee-brown; long, fluffy hairs. **Leaves** 6–22 mm, the shortest ones on top (where a few are held randomly upside-down); broadest (2 mm) *midway*

up; two bright white bands beneath. Aroma of the unrelated poisonous herb Hemlock (sour, parsley-like). **Cones** Drooping from shoot-tips, to 25 mm.

COMPARE Eastern Hemlock (below). Mountain Hemlock: leaves *greyish all round.* Douglas Fir 'Fretsii' (p.66): leaves of more even length carried on green/purple shoots.

Eastern Hemlock 

Tsuga canadensis

(Canadian Hemlock) E North America: Ontario to Alabama. 1736. Now occasional as an older park tree.

APPEARANCE Shape Broad, to 30 m, seldom making a good spire; leader droops only slightly; trunk *sinuous or much forked.* **Bark** Greyer and becoming more shaggily fissured than Western Hemlock's. **Leaves** Like Western Hemlock's, but the shoot-hairs are shorter and the leaves (lemon-scented) are often *broadest near the base* then taper evenly, while a row of very short leaves lies *upside-down on top of each shoot,* their silver undersides conspicuous. **Cones** Small: 18 mm.

COMPARE Western Hemlock (opposite). Other rarer species in cultivation are also often bushy, but lack the regular line of inverted leaves.

VARIANTS 'Pendula' grows a splendid dense dome of weeping foliage; rather rare. Small forms in collections include 'Fremdii' (blackish, glossy, crowded 9 mm leaves), 'Taxifolia', (long leaves crowded at shoot-tips), and the yellow, compact 'Aurea'. 'Microphylla' has very small leaves (6 mm) and 'Macrophylla' largish, crowded ones.

Mountain Hemlock 

Hesperopeuce mertensiana

(*Tsuga mertensiana*) NW North America, in high mountains. 1854. The jumbled needles recall the true hemlocks (or cedar's extension-shoots), while the cones are more spruce-like. Rather rare: vigorous (to 36 m) only in Highland Scotland, but grown elsewhere for its svelte habit and colour.

APPEARANCE Shape Slender-conic, especially when not thriving, or sometimes bending/forking; *separate drooping plates of foliage.* **Bark** Ruggedly scaly; chocolate-brown – contrasting with dark *grey leaves.* **Shoots** Pale brown, shining, with some long hairs. **Leaves** *Irregularly held on either side of shoot,* to 20 mm; narrow, thick and *grey all round.* **Cones** *To 8 cm*; flimsy, downy scales.

DOUGLAS FIR

Douglas Firs (five species) have foliage most like that of the Silver Firs (but the 'suckers' are tiny and only evident when a leaf is pulled off). Their buds are long-conic. (Family: Pinaceae.)

Douglas Fir

Pseudotsuga menziesii

(*P. taxifolia, P. douglasii*) W North America. 1827. The coastal type was once the world's tallest conifer, with trees of 120 m – all now felled for their timber. Abundant away from dry, polluted parts in parks and plantations; rarely naturalizing; Britain's tallest trees now 65 m in Welsh and Scottish mountains.
APPEARANCE **Shape** Spire-like in shelter, on a flag-staff bole. The wispy leader is easily blown out, so old trees often become broad, with heavy, drooping masses of foliage; a few are very weeping. Blackish green; occasionally steel-grey. **Bark** Grey and smooth in youth, then with wide, orange fissures; finally *black- or grey-purple*, and *massively craggy*. **Shoots** Slender, grey-brown, finely hairy. **Buds** Very *slender*, pale brown: almost beech-like. **Leaves** *Soft, flexible and slender* (unlike any silver fir's), to 3 cm, with a *narrowly rounded* tip and narrow, white-green bands beneath; they spread all round most shoots, though with few below. Strong, hot, sweet, fruity aroma fills the air. **Cones** Dropping when ripe; 6 cm. *Three-pronged snake's-tongue bracts*, 15 mm long, point *towards the tip*, but soon break off on the ground.
VARIANTS Blue Douglas Fir, var. *glauca* (E Rocky Mountains, 1876) is healthier than the type (and as frequent as it) in dry lowland areas: a slender, lightly branched, blackish/grey tree, to 30 m only. Bark dark *fawn-grey; scalier* and only shallowly fissured; leaves thicker, less aromatic; often greyish above, with greyer bands beneath; cones small, the bracts *bent to the horizontal* – the most reliable distinction, odd type trees being very blue.

'Fretsii' is a semi-dwarf with very *short, broad leaves* (10 × 3 mm), brightly banded beneath (suggesting a hemlock's, but of even length); 'Brevifolia' has *short* but slenderer leaves, some *curving backwards*. Both are in a few collections.

old tree in exposure

heavy weeping foliage

seedling

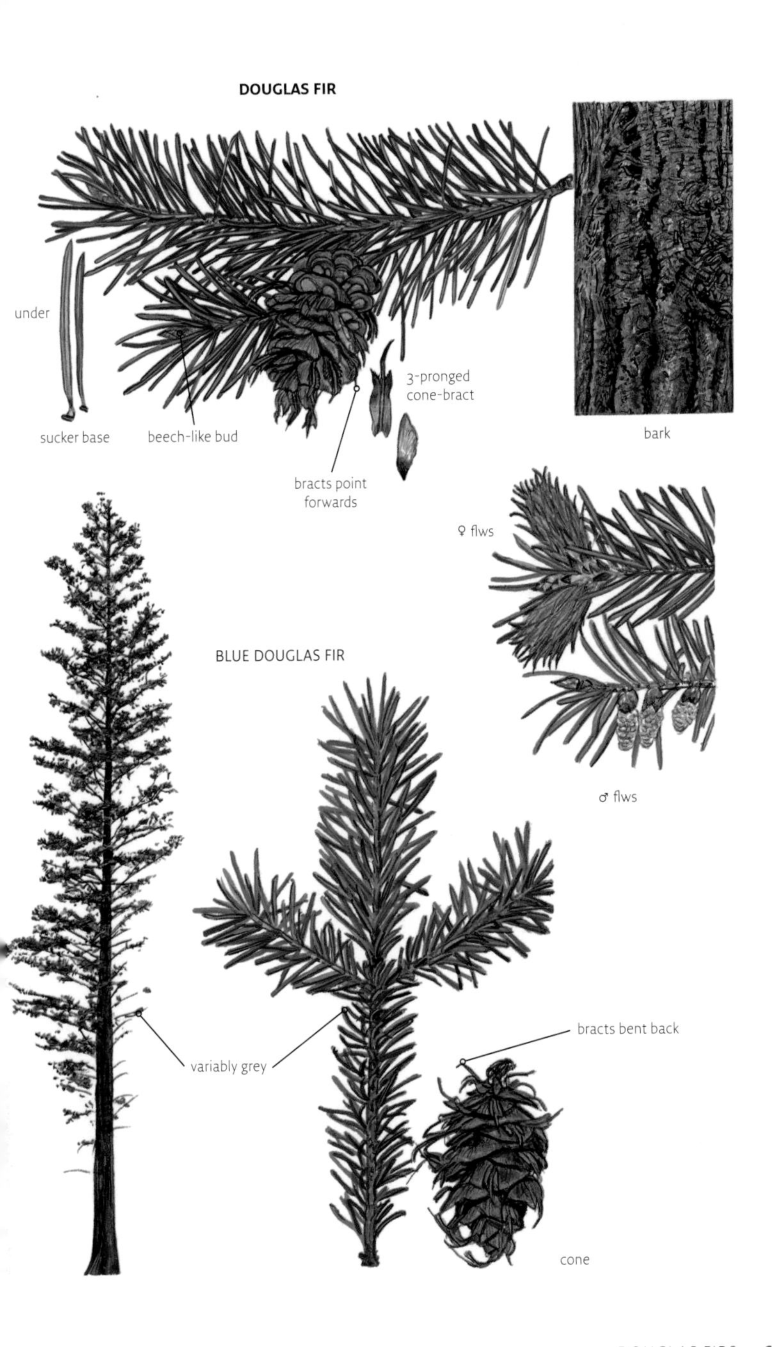
DOUGLAS FIR
under
sucker base
beech-like bud
3-pronged
cone-bract
bark
bracts point
forwards
♀ flws
BLUE DOUGLAS FIR
♂ flws
bracts bent back
variably grey
cone

PINES

Of the pines grown in Britain, all except Singleleaf Nut Pine (*Pinus monophylla*) have adult leaves split into two to eight needles; the number of needles per bundle offers the easiest way to subdivide these trees. The needles are bound by a papery sheath, and juvenile leaves (saplings; sprouts) are undivided. Cones are usually woody, but more spruce-like in the 'soft pines'. (Family: Pinaceae.)

Things to Look for: Pines

- Shape? Bark?
- Buds (if two-needled): Are the scale-tips free? Do they curl outwards?
- Shoots (if five-needled): Colour? Bloom? Hairs?
- Leaves: In twos, threes, fives, or combinations? How long? Twisted (if in twos)? Drooping, or serrated, or white-banded (if in fives)?
- Cones: How big? Spiny? With scales (if five-needled) curving inwards or outwards?

Key Species ✪

Five-needle pines: 'Soft pines' have silky leaves, white-lined on their inner surfaces. Leaves of **Bhutan Pine** (p.75) are long enough (12–20 cm) to droop; **Weymouth Pine's** (p.75) are shorter and held stiffly. **Three-needle pines: Ponderosa Pine** (p.77): long, stiff leaves. **Monterey Pine** (p.76): bright, slenderer/shorter leaves. **Two-needle pines: Scots Pine** (below): short often grey leaves. **Lodgepole Pine** (p.70): short deep-green leaves. **Corsican Pine** (opposite): long leaves; buds with appressed scale-tips. **Maritime Pine** (p.73): long leaves; buds with bent-back scale-tips.

Scots Pine ✪

Pinus sylvestris

Europe (Spain to the Caucasus; Lapland to Siberia), in southern mountains and northern heaths. Scattered natural forests remain in the Scottish Highlands (var. *scotica*); abundantly planted throughout Britain for some 300 years. Common forestry tree; 'weed' on lowland heaths, spreading from old way-marking roundels; yellowish or moribund on chalk. The contrast of soft grey foliage and orange-pink bark makes it one of the prettiest pines.

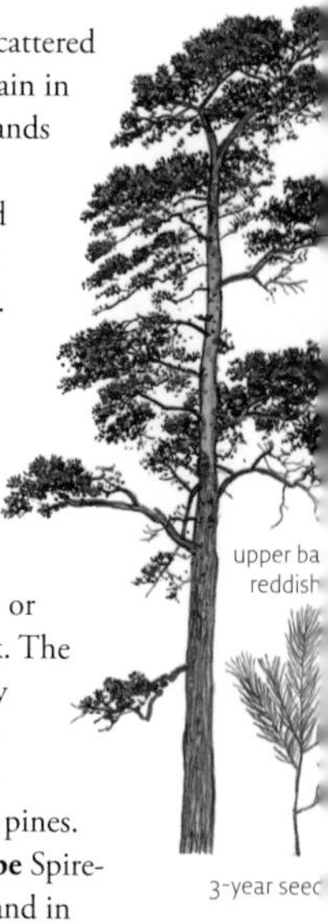

APPEARANCE Shape Spire-like in plantations and in the Baltic var. *rigensis* (preferred for ships' masts; once much planted in Brittany); open-grown trees (especially var. *scotica*) soon picturesquely rounded; to 40 m. **Bark** Red-grey scales at first. The *papery orange-pink bark* intensifies with age in the top half of the tree, while the lower trunk grows big papery-surfaced mauve plates, or sometimes rugged purple ridges. **Shoots** Clear green-brown; hairless. **Buds** With some papery-white scales just free at their tips. **Leaves** In twos, short (5–7 cm), *thicker and often more twisted* than other two-needle pines' except Lodgepole Pine (p.70); pale *blue-grey* in var. *scotica*; blackish and very short in var.

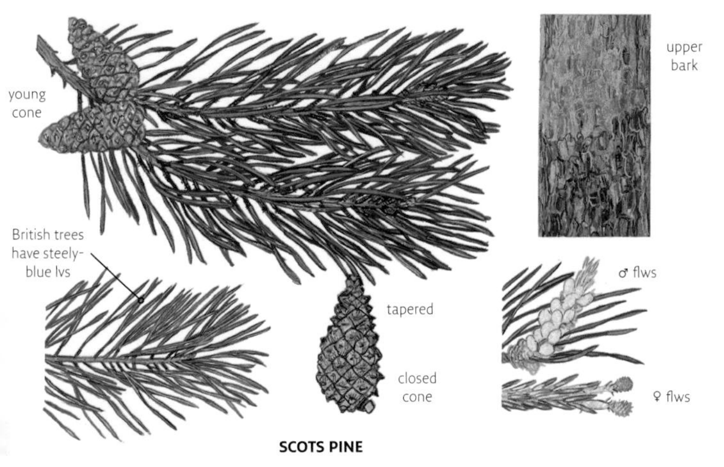

SCOTS PINE

'AUREA' 'FASTIGIATA'

engadinensis (the Alps) and var. *lapponica* (N Scandinavia). **Cones** Slim; 5–8 cm.

COMPARE Many rarer pines have leaves as short but deeper green, and a dull grey bark (see Lodgepole Pine, p.70). Maritime Pine (p.73): often confused, but long leaves and bark are distinct.

VARIANTS All retain orange-pink flaking bark down to a graft-point:

'Aurea' has pale gold leaves in winter; fading a curious pale sickly colour as the foliage emerges grey. Rare.

'Fastigiata' is very slender, with erect branching – in silhouette like a miniature Lombardy Poplar. Rare.

'Watereri' (rather rare) has *very dense short needles* in a picturesque bonsai-like dome (eventually to 15 m), on twisted limbs.

Corsican Pine

Pinus nigra ssp. *laricio*

(*P. n.* var. *maritima*/var. *corsicana*) Corsica; Calabria; Sicily. 1759. Abundant: forestry plantations, shelterbelts, parks.

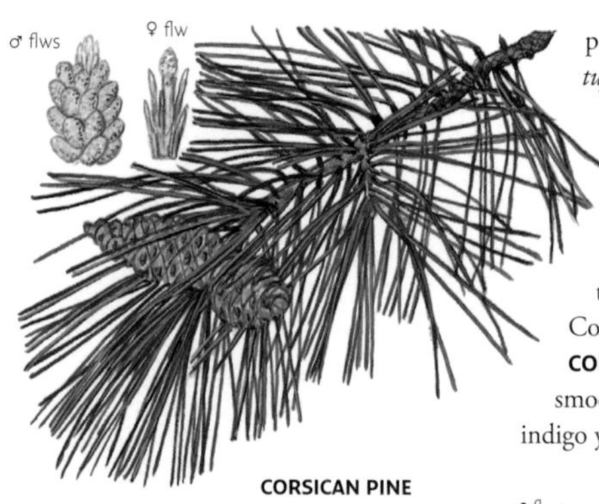

CORSICAN PINE

pointed or tabular; to 43 m. *Dense, tufty foliage, hard, blackish green.* **Bark** *Darker* than Corsican Pine's; the plates often lift shaggily. **Buds** With papery scales, loose but gummed with resin. **Leaves** Shorter (8–14 cm) and thicker than Corsican Pine's. **Cones** As Corsican Pine.
COMPARE Bosnian Pine (p.73): smoother grey bark; leaves to 9 cm; indigo young cones.

APPEARANCE Shape Typically *a straight, clean stem*, often unforked to great heights (45 m); *light* level branches carry quite delicate plates of soft, grey-green foliage, never very dense. **Bark** *Grey*-mauve; big, finely scaly plates and wide fissures. **Shoots** Stout, pale yellow-brown. **Buds** Lacking free scale-tips. **Leaves** In twos, 12–18 cm, rather twisted; slender, greyish. **Cones** Conic, to 8 cm; dull grey-brown.
COMPARE Other subspecies. Austrian Pine (below) is distinguishable by its jizz, but the little-planted Crimean Pine (ssp. *pallasiana*) and Pyrenean Pine (ssp. *salzmannii*) can show intermediate features. Maritime Pine (p.73): reddish, closely plated bark on a sinuous trunk; stiffer leaves; free bud-scales. Ponderosa Pine (p.77): longer leaves *in threes*; stiffer habit.

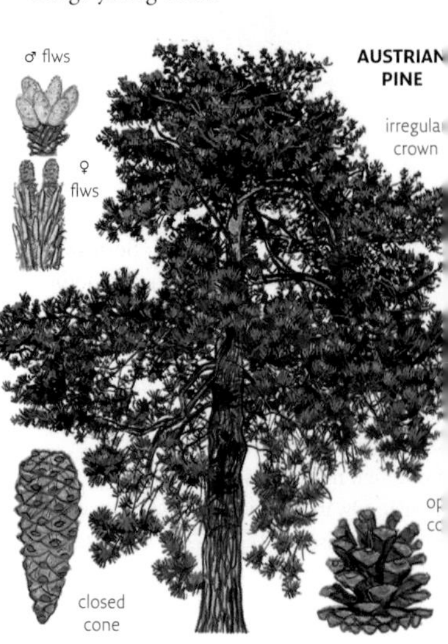

AUSTRIAN PINE

Austrian Pine

Pinus nigra ssp. *nigra*

(Black Pine; *P. n.* var. *austriaca*) S Austria to central Italy and the Balkans. 1835. Very frequent: old parks and shelterbelts. A rough dark pine, but excelling in coastal exposure and on chalk.
APPEARANCE Shape Trunk often long but *seldom quite vertical*; snaggy and generally *heavily limbed*; open-grown trees often broad and much forked from the base,

Lodgepole Pine

Pinus contorta ssp. *latifolia*

Rocky Mountains, from W Alaska to Colorado, where it colonizes in the wake of forest fires. 1854. Abundant in upland plantations; rare in gardens.
APPEARANCE Shape A neat, narrow, open spire to 28 m with small, rising branches, though the trunk often forks. **Bark** Red-brown, *finely scaly*; sometimes very shallowly square-cracked. **Shoots** Green-brown, hairless and glossy. **Buds** With smoothly appressed scales, encased in resin.

thin crown

young tree

Leaves In pairs, 6–10 cm, *broad* and twisted; bright dark green; the needle-sheath is about 5 mm long. **Cones** Small and slender (50 × 25 mm); a minute prickle wears off each scale.
COMPARE Other subspecies. Pines with short bright green or blackish-green leaves in twos include Bosnian Pine (p.73), and Alpine and Scandinavian forms of Scots Pine (p.68). Bosnian Pine has smoother grey bark and indigo immature cones. The upper bark of Scots Pines are orange-pink. Austrian Pine (opposite) has longer (8–14 cm) leaves.
OTHER TREES Mountain Pine, *P. mugo*, from high mountains in S Europe, is similar. Dwarf selections are occasional in small gardens; the western ssp. *uncinata* (*P. uncinata*; rare) can reach 20 m on a straight bole and differs in its very grey, finely-scaling bark and *curiously 'stretched'* lower cone-scales.

Shore Pine

Pinus contorta ssp. *contorta*

Near the Pacific from S Alaska to N California. 1855. Rather occasional everywhere in gardens and belts; in many upland and a few southern plantations.
APPEARANCE Shape Dense, often bushy but sometimes to 32 m, and rounded with age; vivid, blackish green; shoots sometimes twisted. **Bark** Black-brown, soon knobbly with very *small square plates*. **Buds** Often *twisted* as they expand in spring. **Leaves** Shorter (4–6 cm), denser and pressed closer to the shoots than Lodgepole Pine's.
OTHER TREES ssp. *bolanderi* (Mendocino plains, California), is a vigorous form in a

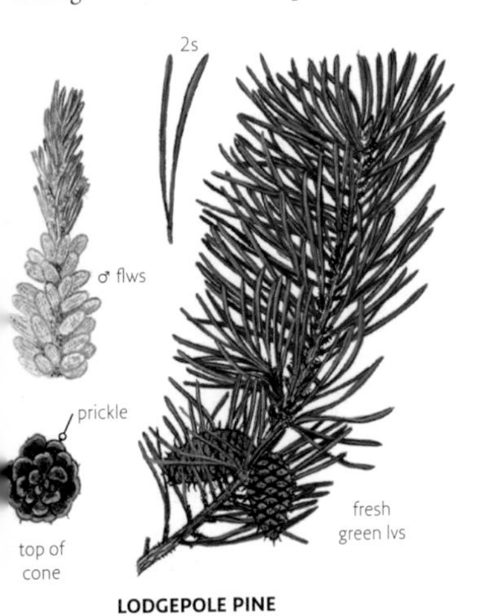

LODGEPOLE PINE

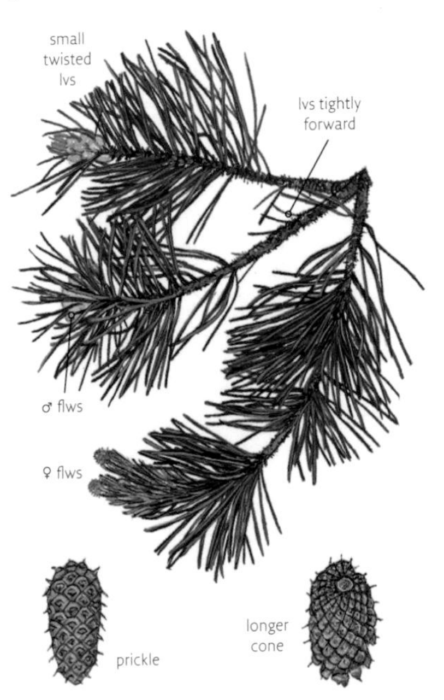

SHORE PINE

few collections: slenderer leaves (*none* with a resin-canal – never a white spot appearing when they are snapped, as there will be for some leaves of the other forms).

Stone Pine 
Pinus pinea

(Umbrella Pine – cf. the unrelated Japanese Umbrella Pine).
Mediterranean Europe; Black Sea coasts of Turkey; long grown in warmer parts of Britain. Very occasional, but in many small gardens, often raised from cones collected on holiday. Saplings, like those of Aleppo Pine, retain single, juvenile leaves for 4–6 years.
APPEARANCE Shape *Many equal leaders* compete by 5 m; rather open, dull and twiggy in shade but in full sun growing a dense, wide *parasol of foliage* (to 20 m) on bare, much-forking limbs which radiate straightly from the often short trunk. **Bark** Deeply fissured from the first and then, at its best, developing *big*, crisp, flat orange-purple plates; in N Europe often darker, greyer and remaining closely fissured. **Shoots** Orange-green. **Buds** *Free, curling scale-tips*, matted together with silvery whiskers (cf. Maritime Pine, opposite). **Leaves** In rather distant pairs, to 16 cm; stiff, straight and dark grey-green, with an oniony scent. **Cones** 10 cm, *very broad, like a fist*, the large seeds a delicacy.

COMPARE Maritime Pine (opposite): taller growth on an often longer, sinuous trunk, longer paler leaves, and slender cones. Scots Pine (p.68): sometimes confused, but a very different tree. Old thriving Stone Pines are unmistakable; the thinner, sparser crowns of struggling examples in N Europe may go unrecognized, but the many joint leaders are characteristic.

Maritime Pine

Pinus pinaster

(Cluster Pine; *P. maritima*) Mediterranean coasts from Portugal to Greece, and Morocco. The source of turpentine and rosin. Long grown in Britain; occasional but locally frequent in SW England (plantations behind some dunes); rarely naturalized.

APPEARANCE **Shape** Open, whorled and often leaning; with age, rounded on a typically long sinuous bole, to 30 m – like Scots Pine (p.68) but much more open; greyish. **Bark** Soon purplish and deeply fissured. Old trees may grow beautiful lizard-skin patterns of small *flat, orange-purple plates* and crisp, black fissures. **Shoots** Stout, pale green-brown. **Buds** Upper scale-tips bent outwards and matted with silvery whiskers (cf. Stone Pine). **Leaves** In sparse pairs, *long* (12–25 cm), *stout and stiff; pale grey-green.* **Cones** Persistent; *slender*; 10 cm, *shining brown*: often collected as ornaments.

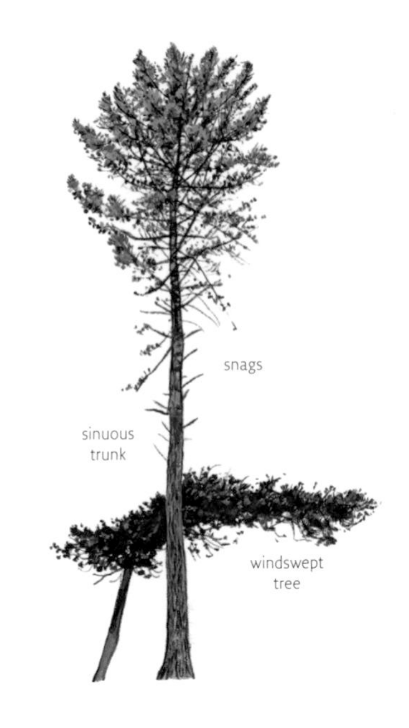

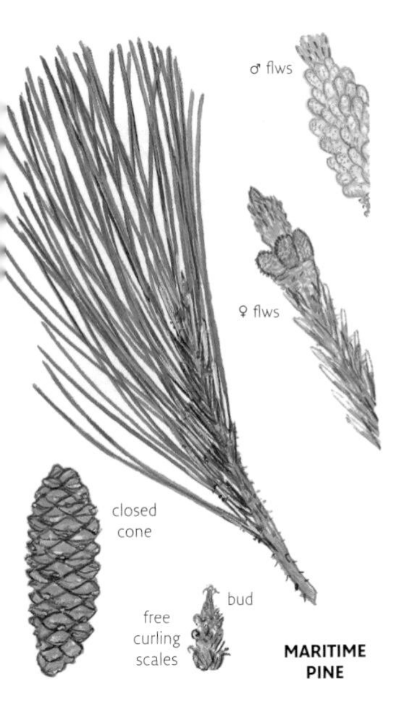

COMPARE Stone Pine (opposite): bark scalier or with longer plates, crown with numerous leaders from youth, cones broad. Bent-back bud-scales also help distinguish from Corsican Pine (p.69), the commonest two-needle pine with long, greyish leaves. Scots Pine (p.68) is sometimes confused.

Bosnian Pine 

Pinus heldreichii

(Incorporating *P. leucodermis*) Balkans; Calabria. 1890. Big gardens, but now much planted.

APPEARANCE **Shape** Often a *neat, vividly Lincoln-green* spire, to 25 m. **Bark** *Ash-grey*; slowly cracking in *neat, shallow squares.* **Shoots** *Grey-bloomed* then *pale* grey-brown. **Buds** Large (15–25 mm); chestnut, abruptly *long-pointed*, the papery scale-tips appressed. **Leaves** In twos, 6–9 cm; stiff and densely forward-angled. **Cones** *Indigo,* finally rufous.

Rocky Mountain Bristlecone Pine

Pinus aristata

W Colorado, N Arizona, and N New Mexico. 1863. In many big gardens: a talking point, though little else. Exceedingly

COMPARE Lodgepole Pine (p.70): yellower leaves, flaky bark, short-pointed buds and unbloomed shoots. Austrian Pine (p.70): longer leaves, rugged bark and dark rough shoots.

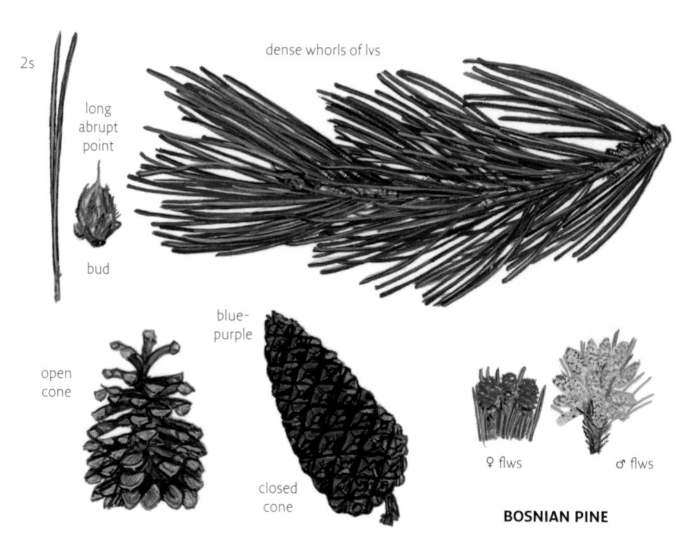

long-lived in the wild on mountains which are too cold and dry for many pests or wood-rotting fungi to survive, though not proving so in the UK. It is thriving on the N coast of Iceland.

APPEARANCE **Shape** Thin, spiky, to 12 m; very slow. **Shoots** Rufous-hairy, little-branched. **Leaves** In fives, *very short* (2–4 cm), incurved, and *densely covering* the shoots for up to 30 years; white bands on their inner surfaces; margins not serrated. Each needle's single resin duct is peculiarly near the surface, and when these break (usually by their second year) the foliage becomes *spotted with white resin, like dandruff.* **Cones** With 5 mm level *spines.*

Bhutan Pine

Pinus wallichiana

(Blue Pine; *P. excelsa, P. griffithii*) Himalayas. 1823. Quite frequent (even in small gardens). Rarely naturalizing.

APPEARANCE **Shape** *Openly* conic, then broad and broken/leaning, to 32 m. **Bark** Purple- or orange-grey; *scaly ridges.* **Shoots** Grey-green, hairless; some *mauve-grey bloom.* **Leaves** Silky, grey-green. In fives, with white-lined inner surfaces; very slender, 10–20 cm – *long enough to droop so most tips lie below shoot-level.* **Cones** Long (10–30 cm); a few small basal scales usually bent back, the rest *straight.*

COMPARE Weymouth Pine (below).

Weymouth Pine

Pinus strobus

E North America: once the region's tallest tree (80 m); but much felled for its fine timber and reduced by Blister-rust. Introduced by Captain George Weymouth in 1705: occasional (in some small gardens); rarely naturalizing.

APPEARANCE **Shape** A spire, soon broad/broken; to 42 m. Crisp *plates of fuzzy foliage.* **Bark** Smooth dark grey, then with rugged but shallow *black* ridges, or redder (cf. Scots Pine's lower bark, p.68).

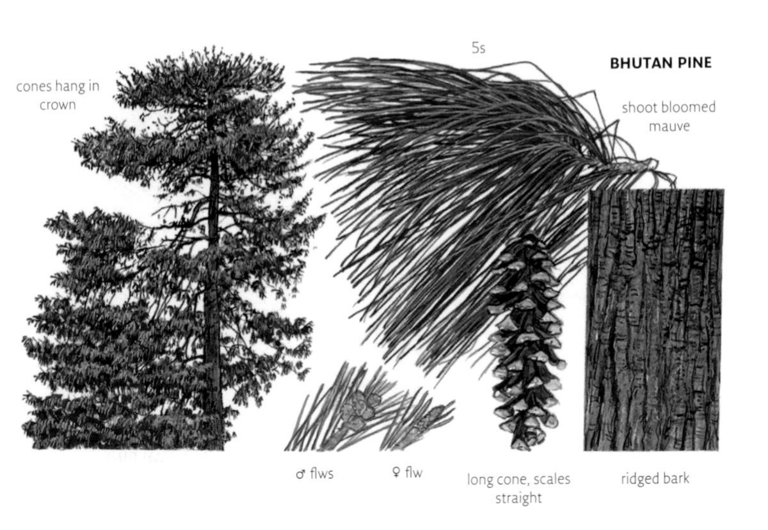

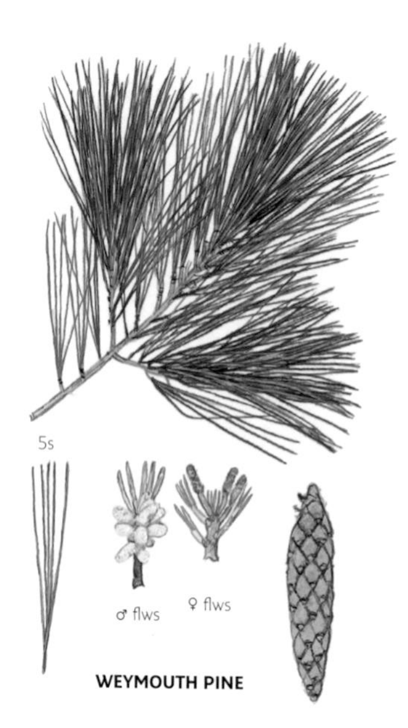

Shoots Slender, green; a transient *tuft* of down behind each leaf-bundle. **Leaves** In fives, with white-lined inner surfaces; 8–12 cm. Slender but short enough to stand *almost straight around the shoot* (forward-angled). **Cones** Seldom to 20 cm, *slender*; a few *small basal scales often curve outwards.* **VARIANTS** 'Contorta' (dense, curved leaves on curiously *wiggling* twigs), 'Fastigiata' (*steep* limbs make a tight spire), and 'Pendula' (*weeping* small branches), among others; all are rather rare.

Monterey Pine

Pinus radiata

The type is confined to three cliffs around Monterey, California: like Monterey Cypress (p.32) it seems to have gone the wrong way when migrating north in its preferred climate zone after the last glaciation, only to find its route cut off by the ocean and by arid coastal zones. 1833. Frequent in milder areas and growing much faster in the UK than in the wild; abundant near some coasts, withstanding salt spray very well; rarely seeding. Occasionally in forestry plantations. (It is New Zealand's principal forestry tree, having reached 60 m in 41 years.)

APPEARANCE
Shape Spikily conic at first, then *densely domed* and sometimes very broad, on heavy, twisting limbs; to 45 m. The bole may be long and straight but is more often short and twisted. The bright green needles appear very dark but brilliant *en masse*. **Bark** Grey; purple-black with age and *very ruggedly fissured.* **Leaves** In threes (but in twos in var. *binata* from Guadalupe Island and var. *cedrosensis* from Cedros Island, Mexico, which have

dark domed crown
broken branches
craggy purplish bark

scarcely been grown in the UK); 10–16 cm long; very slender and nodding. **Cones** Fist-sized, with a minute prickle wearing off each scale; persisting abundantly in whorls and opening only in the wake of forest fires.
OTHER TREES Bishop Pine, *P. muricata*, also from S California (1846; very occasional), can have a remarkably similar jizz but always carried its *thick*, stiffer leaves *in pairs*. Its cones (equally persistent) carry *small prickles*, which may break off with age.

Ponderosa Pine

Pinus ponderosa

(Western Yellow Pine) W Rocky Mountains. 1827. Occasional in parks and gardens everywhere.
APPEARANCE **Shape** A *sparse spire*, often to great heights (40 m); sometimes becoming irregular or rounded but still quite slender. **Bark** At first resembles Corsican Pine's (p.69). Older trees more craggily fissured, blackish or warm red. No distinctive smell. **Shoots** Shiny orange/green. **Buds** Red-brown, resinous; may expand to 5 cm before spring. **Leaves** In threes, long (12–22 cm) and stiff, *like chimney brushes* but sometimes more forward-angled. **Cones** Only *to 15 cm*, a tiny backward-pointing prickle on each scale.
OTHER TREES Jeffrey Pine, *P. jeffreyi*, from similar habitats (1853; less planted), always has a blackish bark whose cracks in summer *smell of wine-gums*; its *large* (to 25 cm) cones have *backward-pointing* prickles. Coulter Pine, *P. coulteri*, from S California into Mexico (1832; rather rare), carries pineapple-sized cones (to 35cm) with *big forward hooks* on each scale; the long needles are particularly sparse.

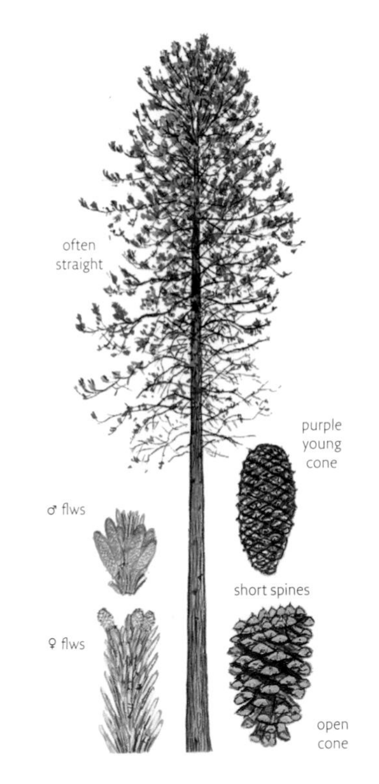

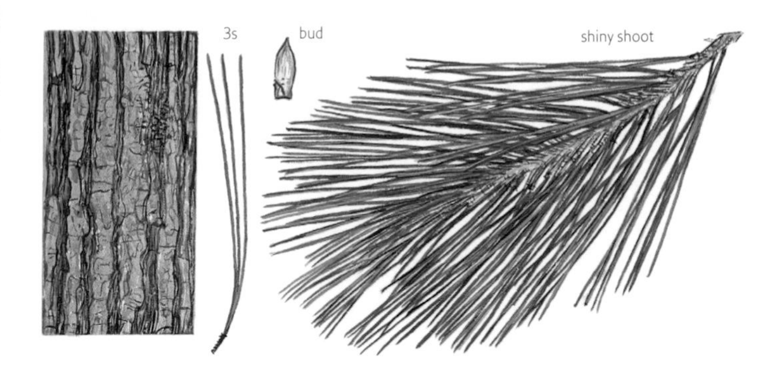

POPLARS

Salicaceae is a family of nearly always dioecious trees and shrubs. Silky hairs on the tiny seeds carry them on the wind; they germinate if they land in wet mud, and all species like a rich, wet soil. Most tree forms suggest a big branch stuck in the ground. The 35 poplars have very sharp buds with many scales, and long, wind-pollinated catkins.

Things to Look for: Black Poplars

- Bark: How rugged?
- Shoots: Are they rising or drooping?
- Leaf-stalk: Glands at top? Hairy?
- Catkins (what gender)?

Key Species: Poplars 

White Poplar (below): lobed leaves, white-hairy under. **Aspen** (p.79): almost round leaves, green under. **Aspen saplings**: small triangular leaves, downy under. **Balsam Spire Poplar** (p.86): leaves smoothly white under. **Wild Black Poplar** (p.80): triangular leaves, green under.

White Poplar ✪

Populus alba

(Abele; *P. nivea*) W and central Eurasia (including Britain?); Tunisia. Frequent. The whitest tree in the landscape.

APPEARANCE Shape Never straight; twiggy and suckering. *Light*, wayward branches; to 28 m only. **Bark** Pale grey in youth, pitted with lines of diamonds (cf. Aspen, p.79; Sallows, p.91). Old trees have creamy-white limbs above black, rugged boles. **Shoots** *White with close wool, which lasts through winter.* **Buds** Dumpy; white-woolly. **Leaves** Unfold furry; hairs rub off the shiny dark grey-green upper side but *plaster the under-leaf all season.* With maple-like lobes on strong shoots; rather rounded on weaker ones (cf. Grey Poplar).

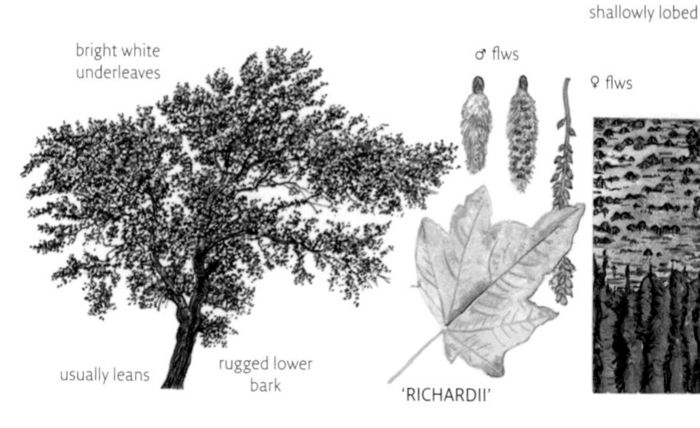

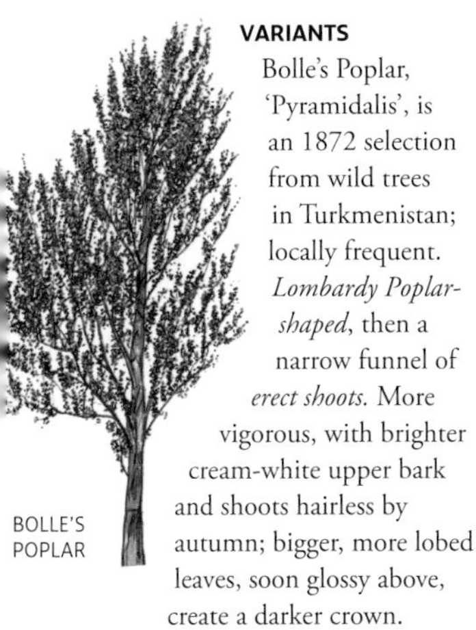

VARIANTS Bolle's Poplar, 'Pyramidalis', is an 1872 selection from wild trees in Turkmenistan; locally frequent. *Lombardy Poplar-shaped*, then a narrow funnel of *erect shoots.* More vigorous, with brighter cream-white upper bark and shoots hairless by autumn; bigger, more lobed leaves, soon glossy above, create a darker crown. Female (green spring catkins). 'Raket' ('Rocket'; by 1956) is a rare spire-shaped improvement; *shoots spreading.*

'Richardii' (1918) has *sunny yellow* leaves (white beneath) – like the type's in autumn. Rare; to 17 m.

Grey Poplar
Populus canescens

(*P.* × *canescens*) Central Europe – a stable hybrid of White Poplar and Aspen. Frequent; naturalized.

APPEARANCE **Shape** *Massive, clean, high branches; to 40 m*; dense *dark* grey-green foliage, jagged/weeping; top branches typically sweep over like a Catherine-wheel. **Bark** Like White Poplar's; base soon very rugged. **Shoots** *Red-grey*; white wool *tends to rub off* by winter. **Leaves** Normally rounded, with big, wave-shaped teeth; more maple-shaped on strong growths. They emerge woolly cream-grey; by summer the older ones are *almost hairless*, the newest still patchily grey-woolly beneath. **Flowers** Most trees are male: purple-grey 4 cm catkins in early spring.

COMPARE White Poplar (above). Can be taken for a giant Aspen (p.79): check for white-hairy young leaves.

VARIANTS Picart's Poplar, 'Macrophylla', has strikingly *large* leaves, to 15 cm wide. Very rare.

Aspen ✪
Populus tremula

W Eurasia (including Britain and Ireland); Algeria. Abundant: suckering stands in scrub, coppices and shingle banks; seldom planted.

APPEARANCE **Shape** *Slender* (exceptionally to 30 m); long bole and *small*, rather perpendicular branches; spiky, gaunt and knobbly with catkin-buds in winter. Soft green; very short-lived. **Bark** Cream, pitted with bands of small black diamonds (cf. White Poplar, opposite; sallows, p.91); then grey and rugged at the base. **Shoots** Shiny

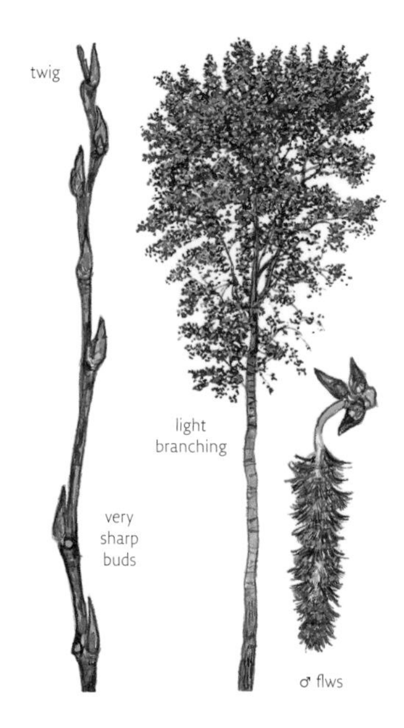

brown on adult tree. **Buds** Long-pointed, *painfully sharp* (like Bird Cherry's, p.193, but soon growing *short side-shoots at right-angles*). **Leaves** *Round,* with big wave-shaped teeth (cf. Grey Poplar, p.79), fluttering (even more than other poplars') in the slightest breeze, so trees are often heard but not seen. Leaves open late and coppery, with grey down, but are *soon hairless.* **Suckers** With triangular green, softly furry leaves on velvety stems: foliage adult by 2 m. **VARIANTS** Weeping Aspen, 'Pendula', has rather stiffly hanging branches. 'Erecta' has *very steep* branches from a straight bole. Both are rare.

Wild Black Poplar ✪

Populus nigra ssp. *betulifolia*

(Downy Black Poplar) NW Europe, including England and Wales. Old trees (all planted?) locally abundant in flood-plains, old parks and some cities but there are only about 6000 in Britain; younger plantings rather occasional – recreation ground shelterbelts, gardens.

APPEARANCE Shape The normally short bole nearly always leans. Soon a huge broad tree to 38 m, of immense presence: heavy, outward-arching limbs with *many small burrs* and *massed, rising shoots*. In summer, richer green and *much leafier*

than its hybrids (p.82–4). Odd trees (accidental crosses with Lombardy Poplars?) have steep, paler grey limbs and more open crowns. **Bark** Greyish *brown* (some old trees nearly black): short deep fissures swirl round burrs and snags. **Shoots** *Amber*, knobbly, with long orange-grey buds. **Leaves** *Small* (7 cm): conspicuously long-tipped, with *no glands at the base*. The young (green) shoots, leaves and stalks have *tiny fine hairs* (cf. the hybrids 'Robusta', p.82, and 'Florence Biondi', p.84), shed by autumn; type trees

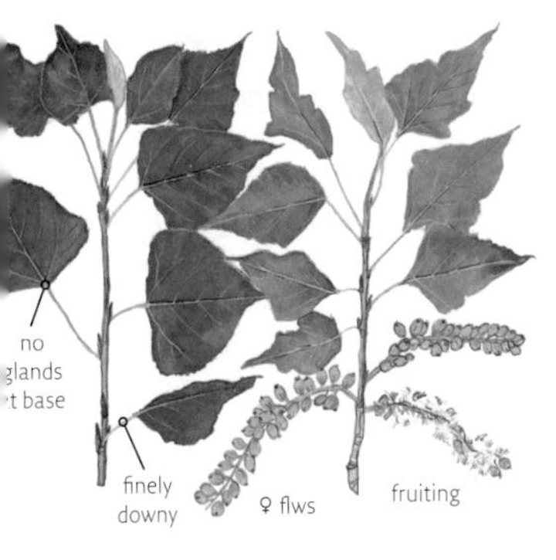

(S/E Europe to central Asia; Tunisia) are hairless. Sweet balsam scent in spring (but milder than in balsam poplars, pp.89–90). **Flowers** Males with red catkins in mid-spring. Old female trees outnumbered by 100:1 ('Manchester Poplar' clones planted in polluted N English cities were all male); green catkins with seed-drop *in May.* **COMPARE** Its cultivars. Railway Poplar (p.82) is much confused.

Lombardy Poplar

Populus nigra 'Italica'

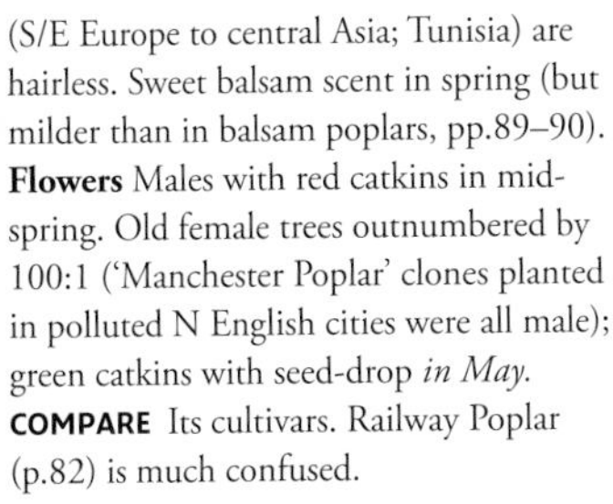

A sport probably of a central Asian race of *P. nigra* (opposite). 1758. Abundant in warmer areas; sometimes naturalizing by suckers.

APPEARANCE Shape Branches remain more tightly vertical than any poplar's except 'Serotina de Selys'; to 38 m. **Bark** Greyish *brown*, moderately craggy and distantly burred with age. **Shoots** Rather *amber; hairless from the first.* **Leaves** *Small, long-pointed,* hairless, in rather dense masses. **Flowers** A *male tree*: red catkins in mid-spring and no 'fluff'. **VARIANTS** Golden Lombardy Poplar, 'Lombardy Gold', is a stumpy sport (to 12 m so far), found in 1974 as a branch of a mature tree in Surrey; very rare.

Plantières Poplar, 'Plantierensis', is a hybrid (Metz, 1884) of the original Lombardy Poplar with the W European Wild Black Poplar (opposite), whose *finely hairy young growths* it inherits: the hairs are best seen on the leaf-stalks and are gone by autumn. Probably frequent in some areas and always worth looking for. Tends to be slightly broader and leafier in maritime climates than typical Lombardy Poplars; the limbs (often sprouty) carry the Wild Black Poplar's *frequent* small burrs. *Stronger balsam smell* in summer (though weaker than Balsam Poplars', p.85–6). Male and female clones exist.

PLANTIÈRES POPLAR 'LOMBARDY GOLD'

Hybrid Black Poplars

Populus × canadensis

(*P. × euramericana*) The cross between the Wild Black Poplar (p.80) and forms of the North American Eastern Cottonwood (*P. deltoides*: now confined to a few collections in Europe) has occurred or been made many times since 1750. With their 'hybrid vigour' the offspring grow furiously, and have been used abundantly in shelterbelts and plantations on rich soil: tending to tower above other trees, they currently dominate most lowland landscapes. The overall appearance (even of known back-crosses with Wild Black Poplars) is much closer to that of the American parent: all develop a *greyer* bark, relatively *regularly* though deeply fissured, and an *airy* crown of constantly susurrating, metallic-green leaves, mid-*green* underneath, *between which the sky often remains visible.* Shoot-tip leaves are very triangular, older ones more wedge-shaped at the base; they are biggest (like other poplars') at the crown-tip, where they drop last in autumn. Young leaves are *fringed with fine hairs,* and at the top of the stalk there are often *one to three small knobbly glands.* Yellowish-*grey* winter shoots carry 1 cm, sharp, narrow buds. Gender is the safest way of distinguishing the several common clones. Females have thin yellow-green catkins in spring which *around mid-summer* shed snowstorms of cotton-wool seeds (the Wild Black Poplar usually sheds its seeds in May); males carry fat red catkins in early spring.

'Robusta' (1895) is now much the most planted clone in plantations and belts; it is also the most distinctive, and so attractive that it earns a place in many gardens.

APPEARANCE **Shape** Straight, rather twiggy trunk, seldom forking; in youth (and to 40 m) roughly *spire-shaped* with regularly whorled, erect branches like a Silver Fir's. The distance between the whorls indicates one summer's growth and can be 2.2 m. Old open-grown trees (few as yet) develop relatively *light,* clean, steep limbs, sometimes from low down, and may lean. **Bark** More *shallowly,* shortly and crookedly fissured than in most clones; paler and a browner grey. **Shoots** Very finely hairy when young. **Leaves** Flush ahead of the other clones, in *mid-spring,* a *pale coppery-red;* rather *large* (10 cm), soon rich green, and relatively *dense,* on stalks minutely hairy at first (cf. 'Florence Biondi' and Wild Black Poplar, p.80). **Flowers** A *male* clone.

COMPARE 'Balsam Spire' (p.86): also spire-shaped and planted everywhere, has a smoother dark grey bark and larger leaves *white* beneath.

'Railway Poplar'. 'Marilandica' (May Poplar) arose around 1800; it has since been submerged in various scarcely distinguishable backcrosses with 'Serotina', known collectively as 'Regenerata' or 'Railway Poplar'. All are *female clones,* with variably dense snowstorms from June to August. Unlike the other hybrid poplars they cope well with exposure, and are locally abundant as mature trees in coastal areas.

APPEARANCE **Shape** Trunk seldom quite straight; often low-forking; typically *more densely domed* than the other clones; to 40 m. Twiggily and roughly conic when grown together (but infrequent in lines and scarce in forestry). Heavy, *outcurving* limbs; many fine shoots *descend.* Sometimes grassy-green summer foliage. **Bark** Grey, often pale or greenish, with deep, generally rather crooked fissures, sometimes some burrs, and often many sprouts. **Leaves** Sparse but *untidily clustered;* they flush often very late, hairless and brownish yellow.

COMPARE Black Italian Poplar and 'Florence Biondi' (p.84); Wild Black Poplar (p.80); Balm of Gilead (p.85).

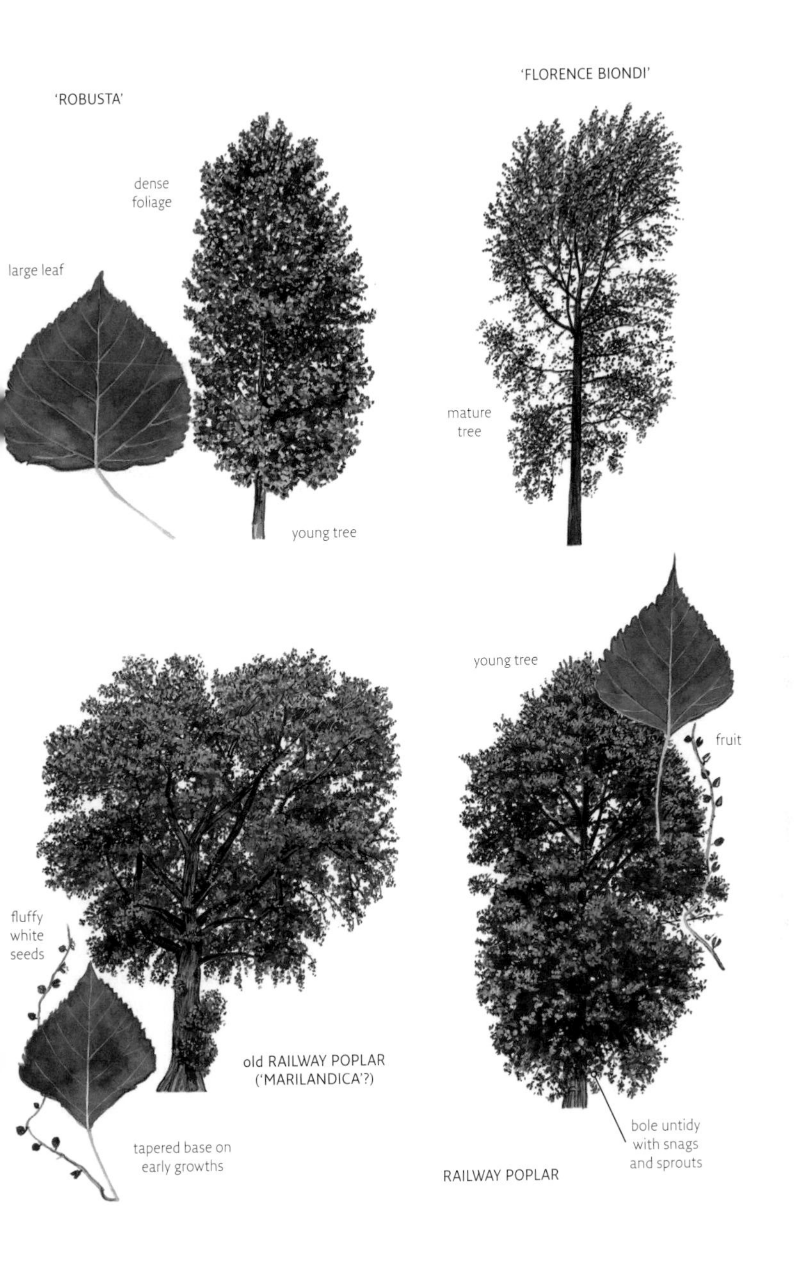
'ROBUSTA'
dense foliage
large leaf
young tree
'FLORENCE BIONDI'
mature tree
young tree
fruit
fluffy white seeds
old RAILWAY POPLAR ('MARILANDICA'?)
tapered base on early growths
bole untidy with snags and sprouts
RAILWAY POPLAR

'Florence Biondi' ('OP226') is a very vigorous, straight-stemmed Hybrid Black Poplar selected by 1950, and now occasional in plantations and belts.
APPEARANCE Shape Cleaner and more graceful than any Railway Poplar clone (p.82). Dark, very sparse foliage; some sprouts make the slightly jagged crown denser along the steeply ascending limbs. **Leaves** Stalks at first minutely downy (cf. only 'Robusta', p.82). **Flowers** A female tree.

Black Italian Poplar, 'Serotina' (France, 1750), features in many tree guides as the common Hybrid Black Poplar, but these clones are short-lived and as this one was little planted in the twentieth century it is now distinctly occasional in most parts.
APPEARANCE Shape A long trunk, seldom straight, typically carries huge, *clean, incurving* limbs; shoots rather thick and ascending. Has reached 45 m. **Bark** Dark/mid grey, soon with long, rather *regular*, very deep fissures. **Leaves** Unfold *very late, a pale coppery brown*; in summer sparse but evenly spread, a dark *sea-green*. **Flowers** A *male clone*.
COMPARE Railway Poplar (p.82).

Golden Poplar, 'Serotina Aurea' (Ghent, 1871), can revert piecemeal to 'Serotina'. (Poplars, growing readily from cuttings, are seldom grafted, so most Golden Poplars cannot be identified from November to June.)
APPEARANCE Shape Domed, twiggy habit, as Railway Poplar (p.82). A locally occasional and eye-catching ornamental: the biggest golden tree (to 30 m). **Leaves** Flush very late and pale brown, then *acid yellow*, finally soft green. (Green clones on chalk can be patchily yellow with chlorosis.)

detail of leaf-glands

Balm of Gilead

Populus x jackii

(Jack's Poplar; *P. candicans*, *P.* × *generosa*) A natural hybrid of Eastern Cottonwood (*P. deltoides*) with Balsam Poplar. In Britain, occasional trees (of both sexes) derive from a cross made at Kew Gardens in 1912.
APPEARANCE Shape At first glance a poor Railway Poplar (p.82), snaggy and often leaning and cankered. **Bark** Pale grey, *less deeply fissured.* **Leaves** Larger, somewhat denser (to 20 cm), *pale whitish green underneath* between a network of green veins; always with two to three glands at top of leaf-stalk.

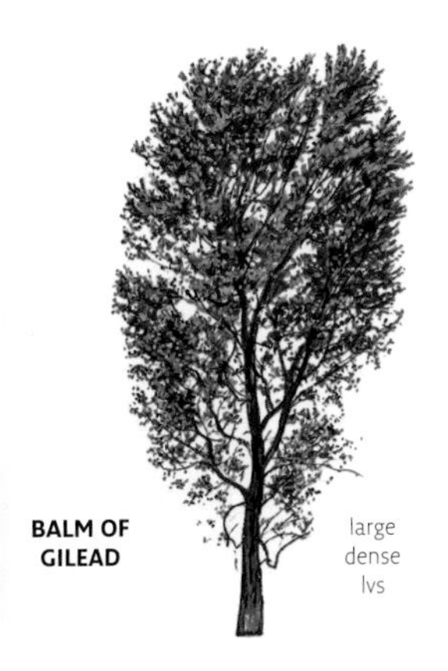

VARIANTS 'Beaupré' and 'Boelare' are recent Belgian selections with big leaves and straight, smooth, pale trunks; hugely vigorous and in some younger plantations.

Variegated Poplar ('Aurora', *c.*1920) is frequent. A snaggy, fragile, much-cankered female tree, to 20 m; unexcitingly dark green in spring (albeit balsam-scented), but until it reverts the big later growth-tip leaves are splashed with, or entirely, cream.'

Western Balsam Poplar

Populus balsamifera ssp. *trichocarpa*

W North America, where it is one of the world's tallest broadleaves (to 70 m). 1892. Now (as var. *hastata*) rather occasional in the UK: susceptible to bacterial canker (*Aplanobacter populi*), short-lived, and largely replaced by its hybrids (below). The sweet scent of the buds' resin, copious in all balsam poplars, *fills the air* through spring.
APPEARANCE Shape Narrow, to 42 m, with steep limbs, soon much broken, on an often densely sprouty bole which usually leans with age; glossily luxuriant foliage *densely clothes* the stems. **Bark** Silvery, or pale brown; shallowly and closely ridged. **Shoots** Red-brown, hairless; slightly *angled* at first. **Buds** Long, sharp and sticky. **Leaves** Variably triangular or oval/heart-shaped, 10–25 cm long; *oily-white underneath between a fine network of green veins* but soon almost hairless; flushing acid-green *early* in spring and briefly yellow in early autumn. **Fruits** Opening into three segments (female trees).
COMPARE Balm of Gilead (above).
VARIANTS 'Fritzi Pauley', a selection from

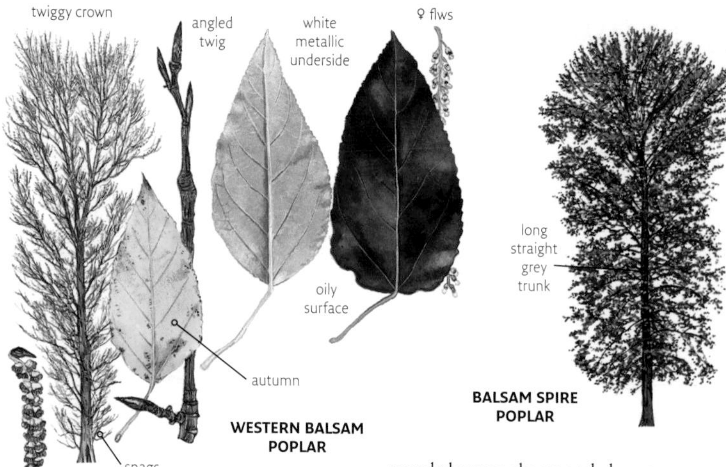

Mt Baker, Washington, is now sometimes used in forestry – especially in cool wet areas – for its canker-resistance, extreme vigour and *straight (but sprouty) trunk* (cf. 'Balsam Spire'), which carries light, rather *level* branches. The bark is *brown*, and soon *rather scaly*; a male clone.

OTHER TREES The eastern subspecies (*P. tacamahaca*; E North America, 1689), now a rarer tree, differs in *prolific suckering* (sometimes naturalized) and in smoothly rounded young shoots and almost untoothed leaves, under which a few hairs *persist*. The clearest distinction (but present only in female plants) is that the seed-pods on the catkins open in two segments, not three.

Balsam Spire Poplar ✪

Populus 'Balsam Spire'

('TT32') The cream of a series of artificial hybrids of Western and Eastern Balsam Poplars, now very abundant: plantations, shelterbelts, parks. Generally canker-resistant.

APPEARANCE Shape At first a dense *spire*; the *straight* trunk seldom forks. Light, *steeply rising* branches make a *spiky fan* at the top of old trees (to 35 m). **Bark** *Silver-black, smooth* for many years, then finely fissured. **Leaves** Shortish, rounded, dark above (but to 30 cm on the strong top-growths). **Flowers** A *female* clone; seed-drop in high summer.

COMPARE 'Robusta' (p.82): a less dense spire, flushing later and coppery-red, with very little balsam scent; 'Fritzi Pauley' (p.85–6); 'Androscoggin'.

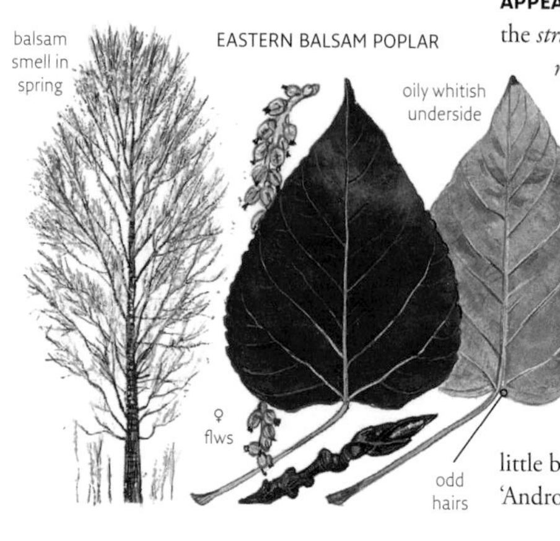

WILLOWS

Willows (400 species, from trees to prostrate sub-shrubs) grow in all continents except Australia and Antarctica. Like poplars, they tend to suggest a big branch stuck in the ground. Their buds have one smooth, flattened scale; the shoots curve evenly and taper to an aborted tip. Short female catkins ('pussies') attract pollinating insects. (Family: Salicaceae.)

THINGS TO LOOK FOR: WILLOWS

- Shape/size?
- Shoots/young bark: Colour?
- Leaves: Downy beneath? How broad? How glossy?

KEY SPECIES 

White Willow (opposite) and Crack Willow (p.89): tall trees with narrow, boat-shaped leaves. **Golden Weeping Willow** (p.92): similar; weeping habit. **Sallows** (p.91): oval leaves, finely grey-woolly beneath. **Bay Willow** (p.89): broadly boat-shaped, hairless leaves. **Common Osier** (p.92): bushy; very narrow leaves.

White Willow ✪

Salix alba

Europe including Britain and Ireland; W Asia, N Africa. Locally abundant but generally confined to the banks of rivers, larger ponds and marshland ditches; the wild form is seldom planted.

APPEARANCE Shape A short leaning bole carries big, *ascending* limbs; the fine shoots *droop*. A large tree, to 30 m, flushing yellow; dark, misty grey in summer, and dull grey-brown in winter; often pollarded. **Bark** Dark *grey*; rugged, criss-crossing ridges. **Shoots** Very slender, grey; softly hairy for a year. **Buds** Slender, flattened, silky. **Leaves**

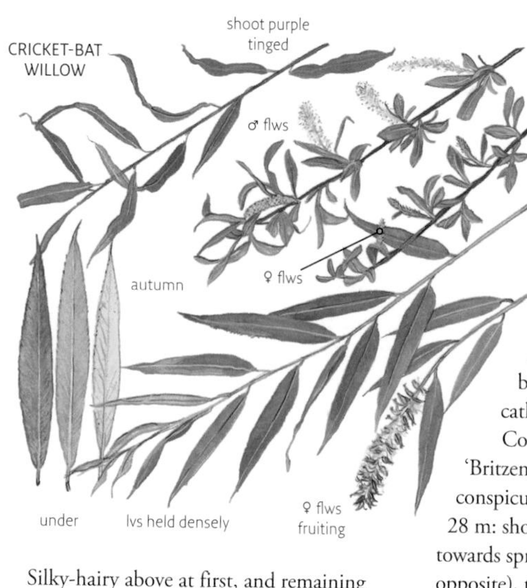

wet ground – growing with extraordinary vigour to 30 m but also providing the best timber for bats; sometimes naturalized. An almost *straight* trunk creates a rough *spire* but can fork low into steep limbs; shoots dark *purplish red* (though finely grey-downy); leaves *almost hairless by late summer* but vivid blue-grey beneath. *Female*: green spring catkins quickly shed fluffy seeds.

Coral-bark Willows (including 'Britzensis' and 'Chermesina') are conspicuous in parks and gardens, to 28 m: shoots *brilliant orange* (amber towards spring; cf. Basford Willow, opposite), the tree in winter sunlight is like a giant flame. Browner bark than the type's; wispily spire-shaped when young, the twigs always *rising and incurving*. Leaves are *soon almost hairless* but blue-grey beneath; crown a distinctive pale *greyish yellow* in summer. *Male trees*, gold in early spring with short catkins. (The female 'Cardinal' is much rarer.)

Golden Willow, var. *vitellina*, is a name sometimes used for Coral-bark Willows but more precisely refers to clones (now rare) with *clear yellow* shoots.

Silky-hairy above at first, and remaining (wild forms) *silver-downy beneath, with odd hairs above*; about 8 cm.

COMPARE Crack Willow (p.89); Willow-leaved Pear (p.178); Oleaster (p.227).

VARIANTS Silver Willow, var. *sericea* (var. *argentea*), is a rather occasional garden tree, whose leaves *stay silky-hairy above*. Slower (to 25 m), with *non-drooping shoots* – like a puff of pale smoke.

Cricket-bat Willow, var. *caerulea*, is frequent, often in lines and plantations in

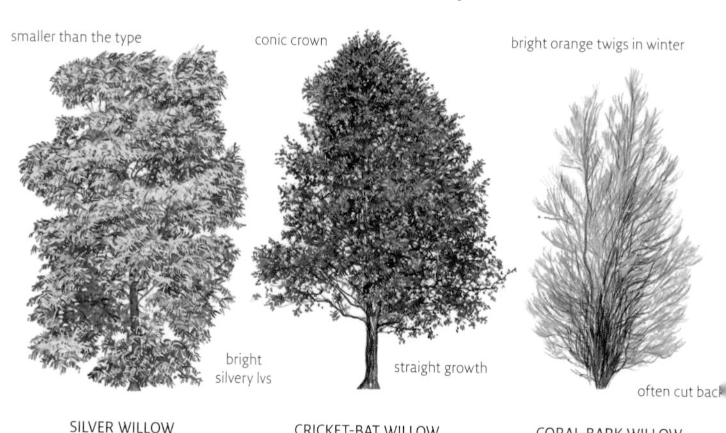

rack Willow ✪
alix × fragilis

urope E to Romania, including Britain nd Ireland. Abundant in wetter ground, ough seldom planted. Originating rehistorically as a hybrid of White Willow).87) with a species from the Caucasus, *S. uxina*. Tall wild trees are likely to be back-rosses with White Willow.
PPEARANCE Shape Broad and short-boled xcept in woodland, with *wide-spreading* ranches and an open interior; twigs not eeping. Glossy-green foliage; dull orange winter (cf. Coral-bark Willow, opposite). eaning and much shattered with age; often ollarded. **Bark** Dark *brown*; very rugged riss-cross ridges. **Shoots** Yellow-brown, niny, soon *hairless*; buds narrow, smooth.

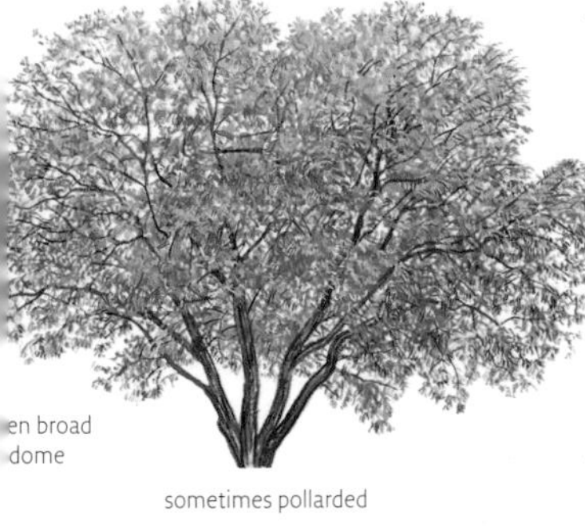

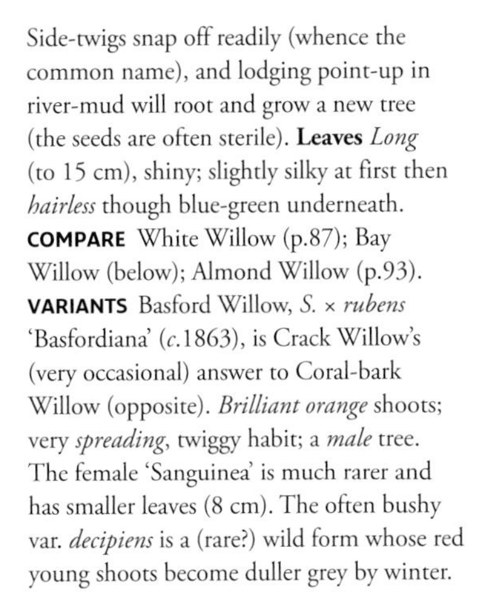

Side-twigs snap off readily (whence the common name), and lodging point-up in river-mud will root and grow a new tree (the seeds are often sterile). **Leaves** *Long* (to 15 cm), shiny; slightly silky at first then *hairless* though blue-green underneath.
COMPARE White Willow (p.87); Bay Willow (below); Almond Willow (p.93).
VARIANTS Basford Willow, *S.* × *rubens* 'Basfordiana' (*c.*1863), is Crack Willow's (very occasional) answer to Coral-bark Willow (opposite). *Brilliant orange* shoots; very *spreading*, twiggy habit; a *male* tree. The female 'Sanguinea' is much rarer and has smaller leaves (8 cm). The often bushy var. *decipiens* is a (rare?) wild form whose red young shoots become duller grey by winter.

Bay Willow ✪
Salix pentandra

N Eurasia (including N England, Scotland, N Ireland and N Wales); in high mountains further S. Frequent by rivers and in wet woodland; in S England a very occasional park tree (rarely naturalized).
APPEARANCE Shape A *dense* dome, to 20 m, or bushy; blackish, almost luminously glossy foliage. **Bark** *Dark grey*; scaly criss-cross ridges. **Shoots** Glossy

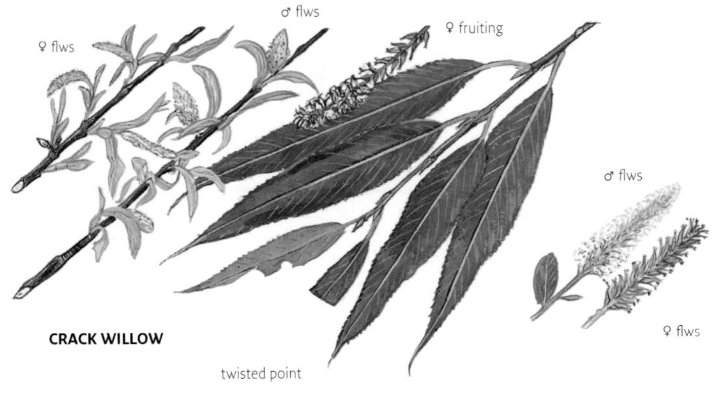

green-brown, hairless. **Buds** Glossy brown. **Leaves** *Hairless*, thick, shiny – *rather like Bay's* (p.150), but deciduous; finely toothed; blue-grey beneath. **Flowers** Catkins on *leafy shoots in late spring* (opening with/before the leaves in most willows).

COMPARE Crack Willow (p.89). The two trees' hybrid, *S.* × *meyeriana*, is intermediate and very rare (as is *S.* × *ehrhartiana* – the hybrid of White Willow, with *odd hairs on both leaf-surfaces*). Almond Willow (p.93): less glossy leaves.

OTHER TREES *S. lucida* (E North America) is much rarer: *long fine leaf-tips*; flowering shoots downy.

Corkscrew Willow

Salix babylonica var. pekinensis 'Tortuosa'

(*S. matsudana* 'Tortuosa') N China/Japan. 1925. Abundant, but very short-lived.

APPEARANCE Shape An electrocuted-looking shock of *corkscrew* branches, then broad-domed and slightly weeping; the fine shoots still curl crazily (cf. Corkscrew Hazel p.110). Pale, soft green until December. **Bark** Pale brown; relatively shallow criss-cross ridges. **Leaves** To 8 cm, variously buckled but not curving in rings. Golden Weeping Willow (p.92) can rarely grow buckled leaves.

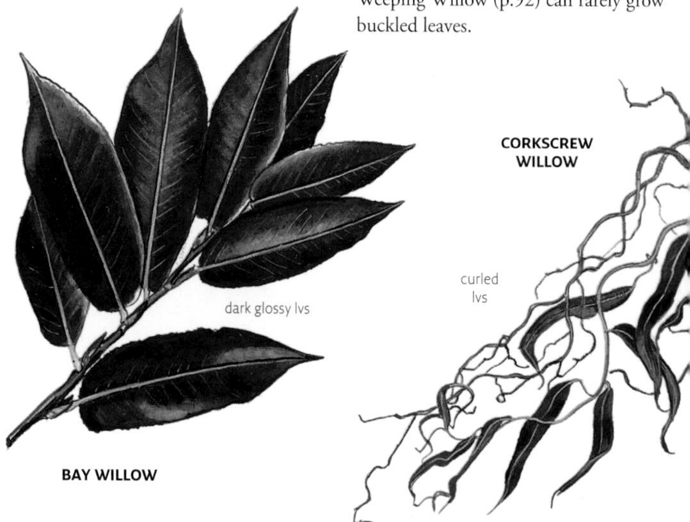

Goat Willow ✪
Salix caprea

(Great Sallow; Pussy Willow) NW Eurasia, including Britain and Ireland. Very abundant except on the lightest soils.
APPEARANCE **Shape** Domed, to 22 m; weak arching branches on usually a single trunk. **Bark** Grey: at first banded with small diamond-shaped pits; soon with rather shallow, criss-cross ridges. **Shoots** Grey (red/yellow in sun), thicker than most willows', *soon hairless.* **Buds** Rather rounded (*downy* in var. *sphacelata* (*S. coaetanea*), from N Europe (Highland Scotland included) and the Alps, which has narrower, almost *untoothed* leaves). **Leaves** *Not more than twice as long as broad,* the *abrupt tip bent sideways*; dark, *wrinkly*; a very fine grey-green felt beneath; few or no teeth. **Flowers** *Precede the leaves.* Male trees' 'pussies' gold; females' silver, quickly shedding fluffy seeds.
COMPARE Grey Sallow (below). A range of rare hybrids (*S.* × *reichardtii*) appear to link the two.
VARIANTS Weeping Sallows, 'Kilmarnock' (male) and 'Weeping Sally' (female), are abundant but very short-lived: weak stems *arch down* from the graft-point (cf. Weeping Osier, p.93).

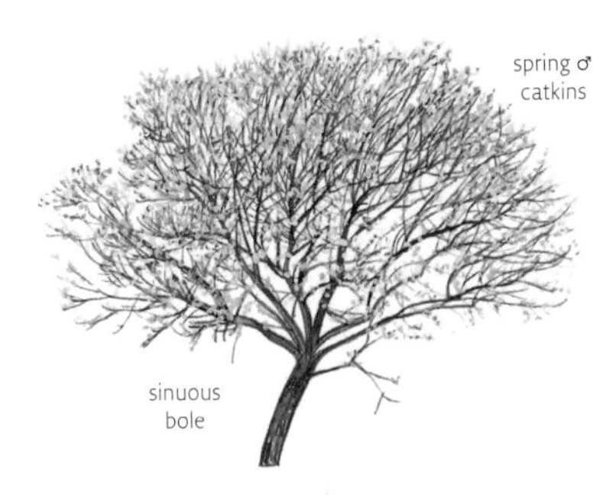

Grey Sallow
Salix cinerea ssp. *oleifolia*

(Rusty Sallow; *S. atrocinerea*) W Europe, including Britain and Ireland. As abundant as Goat Willow, except on dry sites (both are known as 'Common Sallow'); differences as follows.
APPEARANCE **Shape** *Bushy*, seldom long-trunked; to 15 m. **Bark** Grows darker, shallower ridges. **Shoots** Minutely *hairy for*

a year; two-year twigs *ridged* under their bark. **Buds** Minutely *hairy for a year*. **Leaves** Usually much smaller; *2–3 times as long as broad* (cf. Smith's Willow), and broadest above halfway up; a fine felt beneath plus odd *rusty* hairs under the veins. Semicircular stipules are common (as in Almond Willow (opposite). **Flowers** Starting later than Goat Willow's, the 'pussies' slightly smaller.
OTHER TREES The type is local and bushy: downier shoots; greyer leaves lacking rusty hairs.

Golden Weeping Willow ✪

Salix × sepulcralis 'Chrysocoma'

(*Salix alba* 'Tristis'; *S. alba* var. *vitellina pendula*) Berlin. By 1888. Abundant in warm areas, deriving its habit from Chinese Weeping Willow and its vigour and twig-colour from Golden Willow (p.88).
APPEARANCE **Shape** A broad head of twisting limbs (to 24 m) clothed in *long, straight hanging shoots* – 6 m long in ideal conditions. In leaf March–December, *pale yellow-grey-green*. **Bark** *Pale* grey-brown; deep criss-cross ridges. **Shoots** Green, then (in sun) greyish *gold* for many years. **Leaves** Unfolding silky-grey; soon hairless above; hairless beneath in three months (but blue-grey). **Flowers** A male clone (infertile; it grows odd female catkins).

GOLDEN WEEPING WILLOW

Common Osier ✪

Salix viminalis

Mid Europe, including S England. Locally abundant (marshes); much planted/ hybridized for basketry.

APPEARANCE **Shape** Generally bushy, to 10 m: vigorous *wand-like stems* from a short, gnarled, scaly trunk. **Shoots** Soon hairless, yellowish. **Buds** Silky-white, rounded; closely set *like strings of pearls*. **Leaves** *Very narrow* (to 20 × 1 cm); dark and wrinkled above silky-white under; *untoothed* (cf. Oleaster, p.227).
OTHER TREES Hoary Willow, *S. elaeagnos* (central Europe, Asia Minor; in some gardens), has *minutely toothed* leaves, woolly- not silky-hairy beneath.

·lmond Willow

alix triandra

'rench Willow; *S. amygdalina*) Eurasia, ·cluding England, E Wales, S Scotland and Ireland. Once much grown/hybridized for asketry; now rather occasional in marshes/ ɔod-plains.

PPEARANCE Shape Often bushy (rarely › 20 m). **Bark** Grey-brown, *peeling* and aking orange on small branches; later ıggedly ridged. **Shoots** Shiny brown; soon airless. **Leaves** Like Crack Willow's (p.89) ut smaller (5–10 cm) and *hairless*; green or luish beneath. **Male flowers** *three stamens*.

Purple Osier

Salix purpurea

Eurasia, including Britain and Ireland; N Africa. Rather occasional in wet places, including upland bogs. Once much planted/ hybridized for basketry.

APPEARANCE Shape Bushy. **Bark** *Shining grey*, with lenticel-bands; not fissured. Particularly bitter with salicylic acid (the active principle in aspirins). **Shoots** Soon hairless; yellow, or shining *red-purple* in sun (cf. Violet Willow's *bloomed* twigs. **Leaves** Soon hairless; like Crack Willow's (p.89) but smaller (to 12 cm), broadest towards the often blunt tip, and toothed *only* (except in ssp. *lambertiana*) near the apex; *often in opposite pairs*.

VARIANTS Weeping Osier, 'Pendula', is an occasional garden tree whose weak purplish stems arch down from the graft-point (cf. weeping sallows, p.91).

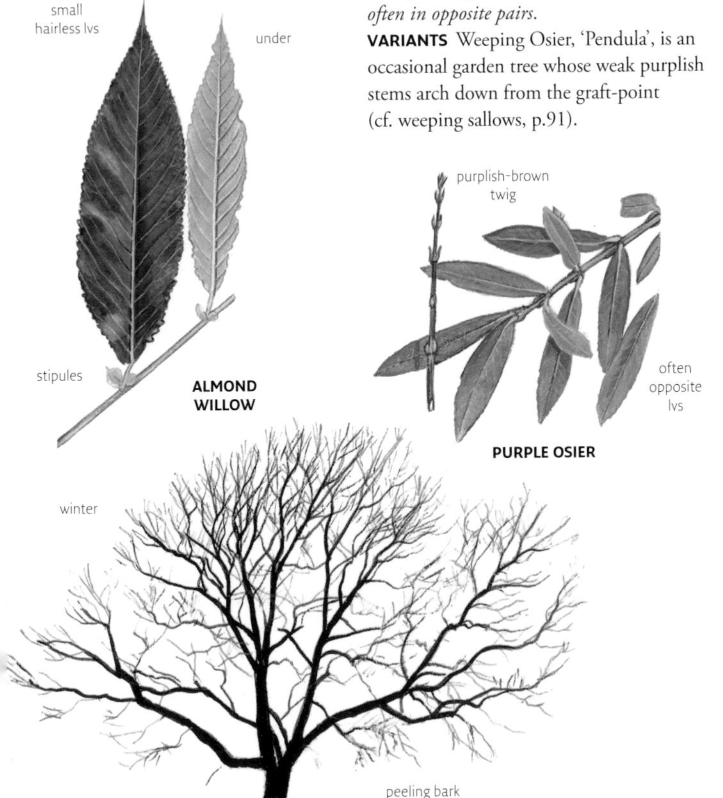

WINGNUT

Wingnut trees (about 8 species) have big, compound, alternate leaves; male (and usually female) flowers in catkins. Like walnuts (but not hickories), wingnuts have a chambered pith – an obliquely cut one-year-old twig shows a series of close divisions (Family: Juglandaceae.)

Caucasian Wingnut

Pterocarya fraxinifolia

Caucasus; N Iran. 1782. Rather occasional, but more recently in many town parks for its almost vandal-proof vigour. **APPEARANCE** Shape Often many-stemmed, or with a cup of heavy limbs from a short lumpy bole. Coarse and gaunt, but magnificent when to 35 m; a vivid, luxuriant green. Trees in unmown areas hide themselves in huge thickets of suckers. Bark Grey-brown: very coarse ridges criss-crossing over one another. Shoots Thick, almost hairless. Buds Often some distance above last year's leaf-scar. No scales – mere miniature leaves, on short stalks and clad in rufous hairs. Leaves Distinguished from those of other trees with large, long, alternate, compound foliage (walnuts, Pecan, sumachs, rowans, Tree of Heaven, Chinese Cedar) by the oblong, stalkless, overlapping, floppy leaflets (up to 25), shiny above and with some pale, long, star-like hairs under the midrib. The central stalk is sometimes finely grooved above but rounded. Autumn colour: clear yellow. Female flowers Catkins, to 50 cm long and conspicuous through summer strung with green nuts which have two 1 cm angled wings.

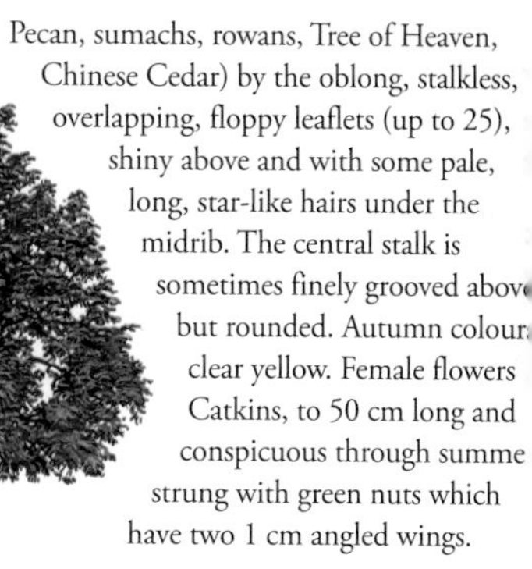

VALNUTS

'alnuts (15 species) have big compound leaves and broadly triangular buds. Fungi sociated with the roots can poison the soil, which limits nearby competition. amily: Juglandaceae.)

HINGS TO LOOK FOR: WALNUTS

Bark
Shoots: Are they hairy or sticky?
Leaves: Are the leaflets serrated?
Fruit: Shape of nuts? How smooth? Are they single or in clusters, or on tails?

ommon Walnut

glans regia

E Europe through the Himalayas and N ırma to SW China. Long cultivated in the K and rarely naturalized on limestone in England; more frequent in countryside ırdens than in towns. It likes a dry climate ıt some of the best of the few big trees e in Scotland. Sometimes long-lived. The mber is the most valuable grown here: trees e dug out rather than felled, as the best ood is at the base.

APPEARANCE Shape Wide and gaunt, on heavy, *twisting*, oak-like branches; to 30 m. Leaves flush late and coppery. **Bark** Shallow, rounded *silvery-grey* ridges; darker grey and rougher with age. **Shoots** Stout, curving, almost *hairless*. **Buds** Squat, purple-brown. **Leaves** With 5–13 (*usually 7*) leaflets: the end one *very large* (to 20 cm), the basal pair *much smaller*. Leaflets oval, *untoothed* (odd peg-teeth on lower pairs and saplings); shiny and *leathery*, hairless except for tufts under vein-joints. Strong smell of shoe-polish. **Fruit** Nuts ripen only in long, hot summers. Many fruiting clones are grown; in orchards, trees traditionally 'distressed' by thrashing the foliage to stimulate fruiting.
VARIANTS Cut-leaved Walnut, 'Laciniata', is rare but attractive: a fresh-green, feathery little tree.

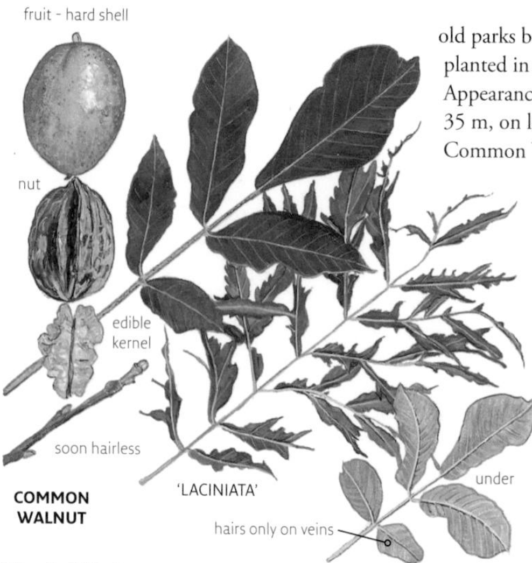

Black Walnut

Juglans nigra

E and central USA; another valuable timber species. Long grown in the UK; rather occasional as a magnificent specimen in old parks but now a little more widely planted in warmer areas.

Appearance **Shape** A grand dome, to 35 m, on less twisting branches than Common Walnut. By early summer a very leafy, rich vivid green. **Bark** Grey or blackish; deep criss-crossing ridges. **Shoots** Brown, hairy; pale grey velvety buds. **Leaves** With 10–23 slender leaflets, but often no end one (cf. Chinese Cedar); leaflets finely toothed and downy beneath, on minutely hairy central stalks. Little aroma. **Fruit** Nuts abundant and strong-flavoured (when, in warm climates, they ripen properly), but in a very thick husk – special nutcrackers are sold in North America to open them. They are carried singly or in pairs except in the clone 'Alburyensis' (original tree at Albury Park,

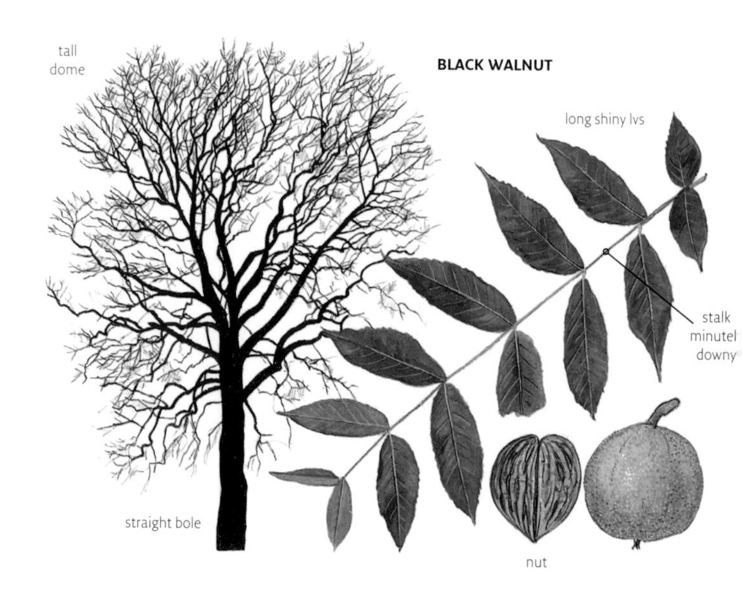

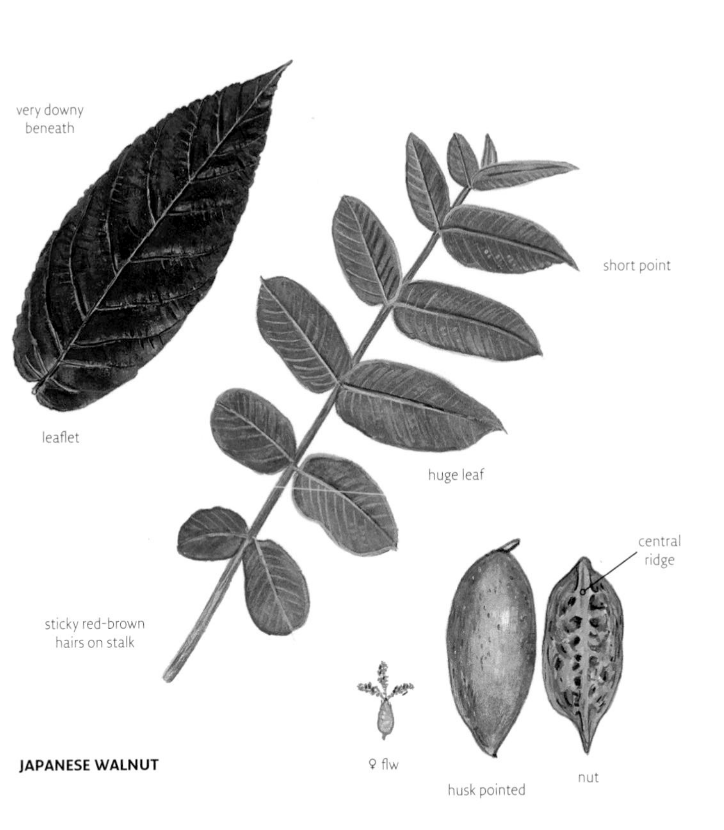

JAPANESE WALNUT

ırrey, with scions in some collections), ʼhich carries clusters of up to five.

OMPARE A very different tree from ˌommon Walnut (above), though hybrids ʼ. × *intermedia*) are known.

apanese Walnut

ıglans ailanthifolia

ʼ. *sieboldiana*) Japan, Sakhalin. 1860. Rare, ut growing well in Scotland.

PPEARANCE **Shape** Broad but sometimes ıunt; to 20 m; suckering. **Bark** Grey, ith shallow ridges: slightly rougher than ommon Walnut's (p.95). **Shoots** With fine hite streaks, and short dense sticky whitish ıirs for up to three years. The big, pale leaf-scars are trefoil-shaped. **Leaves** *Huge* (to 1 m); 9–21 big, rather oblong leaflets, bright and shiny but finely downy above and very downy beneath; they taper *abruptly to a short point*. The central leaf-stalk is covered more densely than Butter-nut's in dark red, sticky down. **Fruit** Nuts *clustered on a long* (15 cm) *hanging tail*, in rinds with sticky hairs; husk pointed, with a rather *smooth surface but a prominent ridge where the halves meet*. In late spring the flower-head is erect and the female flowers (*12–20*) are quietly spectacular – paired, 1 cm crimson plumes. In the Heart-nut (var. *cordiformis*, cultivated in Japan for its fruit, and not found wild), the husk is *flimsy* and heart-shaped at the base.

BIRCHES

Birches (60 species; hybridizing freely) have light airy crowns and typically bright peeling bark. The trees display their male catkins through the winter months. Their shoots are slender, with big, conic, often sticky buds. (Family: Betulaceae.)

Things to Look for: Birches

- Bark: What colours? How rugged?
- Shoots: Are they downy?
- Leaves: How dark/glossy? Are they doubly (regularly) toothed? Is the leaf-stalk downy? How many leaf-veins (how close? how parallel?)?
- Fruiting catkin-scales – Downy? Lumpy?

Silver Birch ✪

Betula pendula

(Warty Birch; *B. verrucosa*; *B. alba* in part) Europe including Britain and Ireland (this and Downy Birch are the only wild trees in Iceland); NW Asia. Transiently dominant on sandy soils. One of the prettiest, airiest birches, planted everywhere.

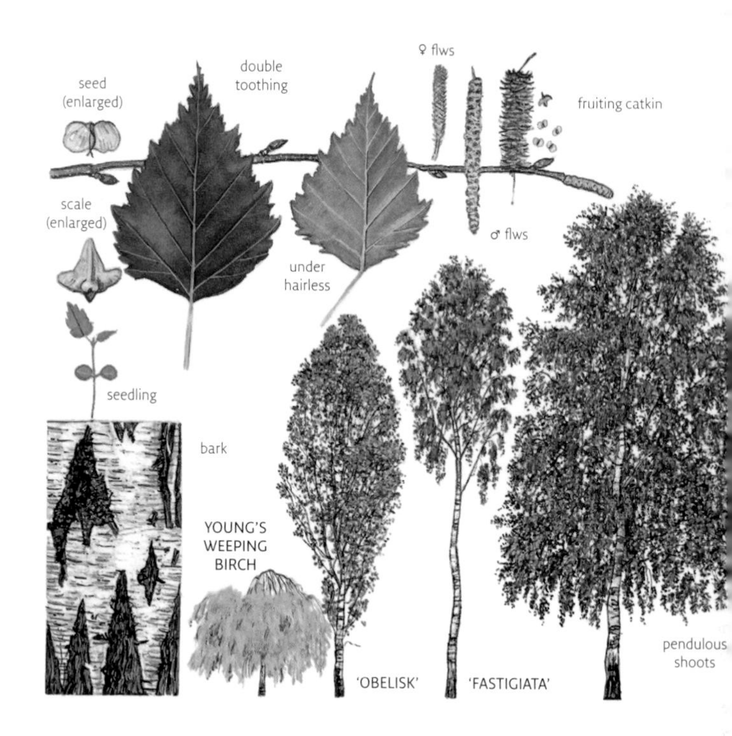

APPEARANCE Shape Twigs soon *weep*, like ountains on the best trees. To 30 m; short-ved except in Highland Scotland. **Bark**)range-red on young stems; soon white, ut growing *rough black arrows/diamonds* ntil the whole butt is *dark and rugged.* **hoots** Hairless (except for strong sprouts); urple-brown; *scrubby with little white warts* specially in sun. **Leaves** *Hairless, on hairless talks*; very triangular: *double teeth up the traight sides.*

OMPARE Downy Birch. (The ybrid, *B.* × *aurata*, is widespread and ntermediate.) Weeping habit distinguishes rom the exotic species, but accidental ntermediate planted hybrids occur.

ARIANTS Young's Weeping Birch, Youngii', is abundant: a *mop of long, anging shoots.* Old trees make some upward rogress and develop extraordinary, fluted unks with big black and white zones, ke a Right Whale's head. 'Tristis' is more ignified, but occasional: as weeping as ne best wild trees; smoothly white-barked bove the graft.

'Golden Cloud' has soft yellowish foliage; ery rare (birches on chalk are often yellow om chlorosis).

'Fastigiata' is rather rare: *vertical wriggling branches*, as if electrocuted, with some drooping shoots, make a *narrow* balloon-shape (cf. Downy Birch). 'Obelisk' is a newer improvement.

Swedish Birch, 'Laciniata' (Cut-leaved Birch; 'Dalecarlica'), is occasional: an airy, shapely grey crown of cut leaves on a smoothly rounded, very white trunk. 'Birkalensis' is in some collections.

OTHER TREES *B. obscura* (E Europe; in some big gardens in the UK) usually has a duller grey bark and dark, rounder leaves, more tapered at the base.

Downy Birch
Betula pubescens

(White Birch, Brown Birch; *B. alba* in part.) Europe (including Britain and Ireland); W Asia. Abundant everywhere on poor or damp non-chalky soils – more widespread than Silver Birch, but less planted.

APPEARANCE Shape Twiggy, to 28 m; the fine branches scarcely weep. **Bark** Purple-red when young, taking longer than Silver Birch's to whiten. Old trunks have *bands of grey*, but little sharp, vertical patterning.

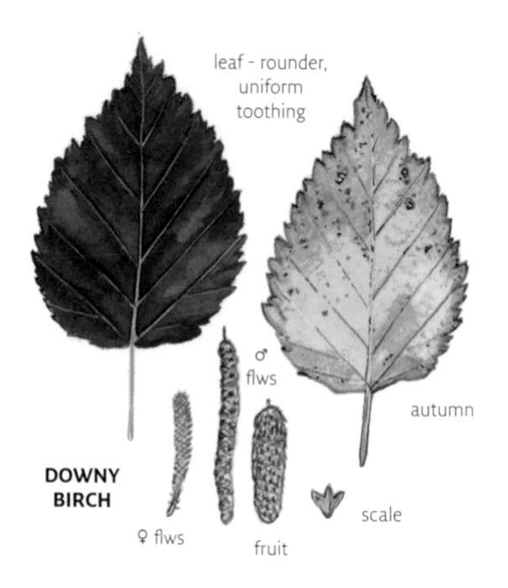

Shoots *Softly hairy* (hairless *but with sticky brown warts* in ssp. *carpatica* from upland regions, including Highland Scotland).
Leaves *Rounded*-triangular, *single-toothed*, on *downy* stalks.
COMPARE Silver Birch (above). Hybridizes freely with exotic species (which usually have bigger leaves).
OTHER TREES *B. celtiberica* (N Iberia; in a few collections) has leaves like Downy Birch's, but on shoots which have Silver Birch's *white* warts.

Paper-bark Birch ✪

Betula papyrifera

(Canoe Birch) Across North America, S to New York. 1750. Locally frequent: gardens, streets.

APPEARANCE Shape Stiffly rising, spreading branches, to 20 m. Lacks much of Silver Birch's grace (p.98): thicker shoots and *sparse*, big, dark leaves. **Bark** Slightly whiter than the European native birches' (but shiny brown in some varieties), peeling horizontally: fine dark lenticel-bands, but rough zones only if the 'paper' is torn off. **Shoots** Warty; some long hairs at first. **Leaves** Large (to 10 cm in some forms), the *disproportionately few main veins* (5–10 pairs) imperfectly parallel. *Du*[ll] dark green; scattered black dots (glands) beneath. *Quite long hairs on the leaf-stalk*, under the vein-joints, and scattered above the leaf.

COMPARE Himalayan Birch forms (p.101): *glossier* leaves with closer veins; hairs extend under the veins. Erman's Birch (p.102): fresher green leaves and *no hairs* on its mature shoots or (usually) its leaf-stalks.

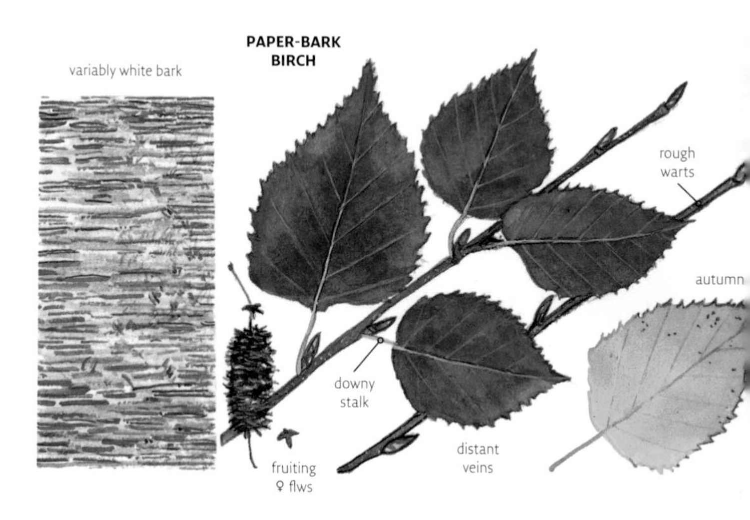

THER TREES Grey Birch, *B. populifolia* North America), is in a few big gardens: naller, hairless, doubly toothed leaves with *nspicuously long, jaggedly toothed tail-like ints*. Blue Birch, *B. coerulea-grandis* (E orth America, 1905; very rare), differs its short-pointed, singly toothed leaves. :iffer habit than Silver Birch, p.98; larger cm) leaves; scarcely patterned white bark.)

iver Birch

etula nigra

lack Birch) E USA. 1736. Occasional, but creasingly planted in parks and streets. **PEARANCE Shape** Broad-limbed with ooping shoots; to 16 m. **Bark** Cream at st, soon with great *scrolled scales*; rufous *almost black* in age. **Shoots** Often downy. **eaves** *Long, with double, scalloped teeth*, iiry at least under the veins and on the alks.

THER TREES *B. davurica* (N China, Korea; 322) is a small tree in some big gardens ith similar bark and less markedly double-othed leaves.

Himalayan Birch

Betula utilis

Himalayas to W China. 1849. Planted for its attractive bark: named selections are increasingly frequent in smaller gardens, parks and streets.

APPEARANCE Shape Somewhat stiffly rising branches to 22 m; quite sparse leaves. **Shoots** Relatively stout; *very hairy* at first. **Bark** Amazing palettes of brilliant, glistening colours, even on 3 cm stems: dead white in the W Himalayan var. *jacquemontii* (*B. jacquemontii*); golden; pale mauve; salmon-pink; crimson or (var. *prattii*) purple. Horizontally marked with small grey/amber lenticels; often with papery, peeling rolls and, rarely, harder curling scales; the white pigment (betulin) may rub off on the hand. **Leaves** *Dark, rather glossy*, on hairy stalks; 5–9 cm; hairs scattered above them, and under the veins which are in 7–14 pairs (most in eastern trees; 7–8 pairs in var. *jacquemontii*). **Flowers** Male catkins in winter have embossed scales.

COMPARE Erman's Birch (opposite): shoots soon hairless. Paper-bark Birch (p.100): *dull* leaves with distant veins, scarcely hairy beneath.

variant barks

var. *PRATTII*

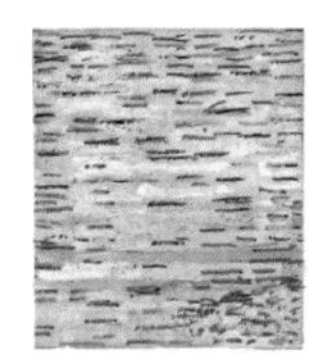

var. *JACQUEMONTII*

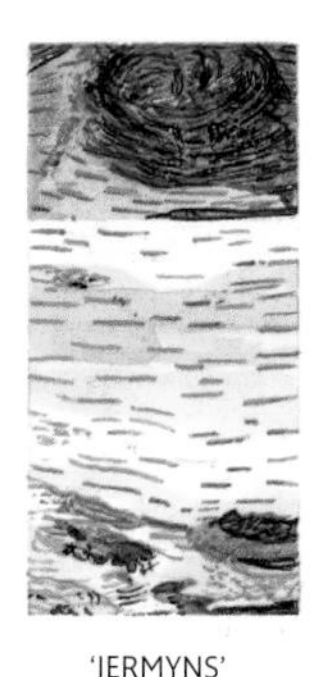

'JERMYNS'

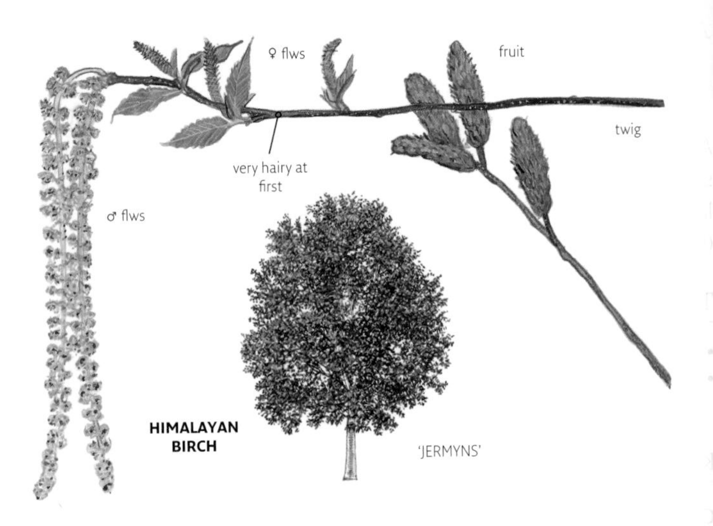

VARIANTS Most named selections ('Grayswood Ghost'; 'Silver Shadow'; 'Jermyns' – neatly steep-branching, with catkins to 17 cm) are intensely white-barked forms of var. *jacquemontii*, often grafted (at the base) on Silver Birch.

Erman's Birch

Betula ermanii

NE Asia, Japan. 1890. Occasional: larger gardens.

APPEARANCE Shape A vigorous, potentially long-lived birch, to 22 m; *dense, often yellowish foliage*. **Bark** Often brilliant *golden*-white, horizontally peeling and *shredding*; sometimes pinkish; rarely shining orange. **Shoots** Warty, *soon hairless*. **Leaves** Triangular/heart-shaped and particularly *neat*; soon almost hairless, with 7–11 often *closely parallel, rather impressed vein-pairs* (14–15 in var. *japonica*); stalks usually hairless. **Flowers** Male catkins in winter fat and *smoothly* scaled.

COMPARE Himalayan Birch (above): hairy shoots. Paper-bark Birch (p.100): hairy shoots and relatively distant, imperfectly parallel leaf-veins.

OTHER TREES *B. costata* (NE Asia) has slender triangular leaves, very *long-pointed* and *never heart-shaped at the base*, with 10–14 vein-pairs. In some collections (though most labelled trees seem to be 'Grayswood', a white-barked clone of *B. ermanii* var. *japonica*). Many older planted 'Erman's Birches' are clearly hybrids with Silver and Downy birches.

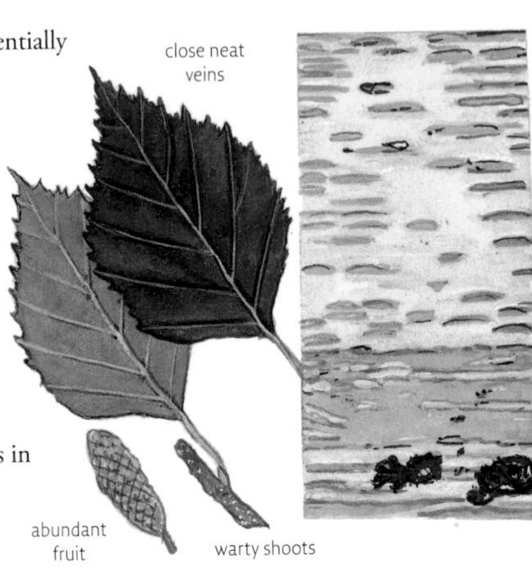

ᴧLDERS

lders (30 species) carry their buds (except in the Green Alder group) on small slim
:alks; male catkins exposed colourfully through winter; female catkins ripening into
ttle woody 'cones'. Their roots develop nodules, enhancing the fertility of poor soils.
'amily: Betulaceae.)

'HINGS TO LOOK FOR: ALDERS

Shoot: Is it hairy?
Leaves: Pointed? Does the rim curl under? Are they downy/glossy beneath?

:EY SPECIES 

:ommon Alder (below): leaves often with ıdented tips. **Italian Alder** (p.105): heart-ıaped leaves. **Grey Alder** (p.105): well-ointed oval leaves.

:ommon Alder

.lnus glutinosa

Black Alder) Europe including Britain nd Ireland; W Asia; N Africa. Abundant but little planted except in reclamation schemes on spoil heaps or landfill sites; often dominant in bogs, 'carrs' and on river-banks. By 2002, many trees had been killed by a new water-borne *Phytophthora* root pathogen.

APPEARANCE Shape An approximate spire when young; old trees (to 28 m) can be broad, with twisting oak-like branches, but in woodlands retain long straight boles. Often coppiced (the timber – white when cut, but oxidizing orange-red in minutes – made ideal charcoal for gunpowder). **Bark** Brown: pale horizontal lenticels, then closely and deeply square-plated, the verticals predominating. **Shoots** Hairless. **Buds** All stalked. Lumpily *club-shaped,* exquisitely *mauve-bloomed* (sometimes dull and greyer).

Leaves Dark, leathery, and *racquet-shaped*; the end never pointed and *often indented* (cf. Alder Buckthorn, p.221). On vigorous growths, they can have very shallow, rounded lobes. **Flowers** Male catkins densely wine-red through winter.
VARIANTS 'Imperialis' (occasional) is a gaunt, 'Japanese-looking' tree with soft green, very *feathery foliage* – with the jizz of Swamp Cypress (p.42).

'Laciniata' (rarer) grows like the type; regular triangular lobes run halfway to the midrib (cf. Wild Service, p.163). The equally rare and similar if rather more elegantly lobed Grey Alder 'Laciniata' (p.105) has *downy shoots* and different bark.

'Pyramidalis' is a very rare steep-branched spire.

'Aurea' is rare. Chlorotic trees of the type may also have patchily yellow foliage; Golden Grey Alder (p.105) is commoner.
OTHER TREES Downy Alder, *A. hirsuta* (Japan, Manchuria; 1879), in some collections and to 20 m, shares racquet-shaped, somewhat lobed leaves, but these are doubly toothed, larger, and variably *rufous-hairy beneath*. Its catkins open in mid-winter.

Red Alder
Alnus rubra

(Oregon Alder; *A. oregona*) S Alaska to California. Occasional; large gardens and a few shelterbelts.
APPEARANCE **Shape** A very vigorous, broad, leafy spire, on light, rising branches. To 24 m; short-lived in Britain. **Bark** Grey; rather smooth. **Shoots** Waxy, angular; long

airs *soon shed.* **Leaves** Large (to 15 cm), eeply double-toothed or shallowly lobed; ark rich green above and grey beneath hough downy only under the impressed eins); the edge is *minutely but sharply rolled own* so that the under-surface is rimmed ith dark green. The stalks are shorter than most alders.
OMPARE Grey Alder: hairier shoots and naller, less lobed leaves (edges not rolled own).

rey Alder ✪
lnus incana

urope except Britain; the Caucasus. 1780. ocally abundant in belts, reclamation hemes, etc.; rarely naturalizing.
PPEARANCE **Shape** Vigorous trees broad-onic; often leaning/stunted and *suckering.* o 24 m: short-lived. **Bark** *Grey*; distantly racked at maturity. **Shoots** *Grey-hairy* hen younger. **Leaves** To 10 cm; often road but always pointed; deeply toothed/ hallowly lobed; dull above, grey and more r less downy beneath.
OMPARE Red Alder (above). The hybrid ith Common Alder, *A.* × *hybrida* (*A.* × *ubescens*), crops up occasionally and is ntermediate.

VARIANTS 'Aurea', rare and slender (to 12 m), has yellow foliage; its yellow shoots turn orange in winter. 'Ramulis Coccineis' (less scarce) is similar in summer; in winter a heart-warming sight – catkins *salmon-pink*, shoots *fiery-red.*

'Pendula', rare and low, weeps rather gauntly.

'Laciniata', rare, has sharply and deeply lobed leaves (cf. Common Alder 'Laciniata', p.103).

Italian Alder ✪
Alnus cordata

S Italy; Corsica; NW Albania. 1820. Locally abundant in parks, streets and shelterbelts. A plant of vigour and polish – even on dry chalk – which still blends well with wild trees.
APPEARANCE **Shape** A spire at first; eventually leaning but usually still narrow; to 28 m. **Bark** Pale brown-grey; vertically plated with age. **Shoots** Bright brown, grey-bloomed. **Leaves** *Dark, glossy, heart-shaped*; like Common Pear's (p.177) but bigger (4–12 cm) – pointed in Italian trees and more rounded in Corsican ones; hairless except for orange tufts under the vein-joints. They flutter on their long stalks. **Flowers**

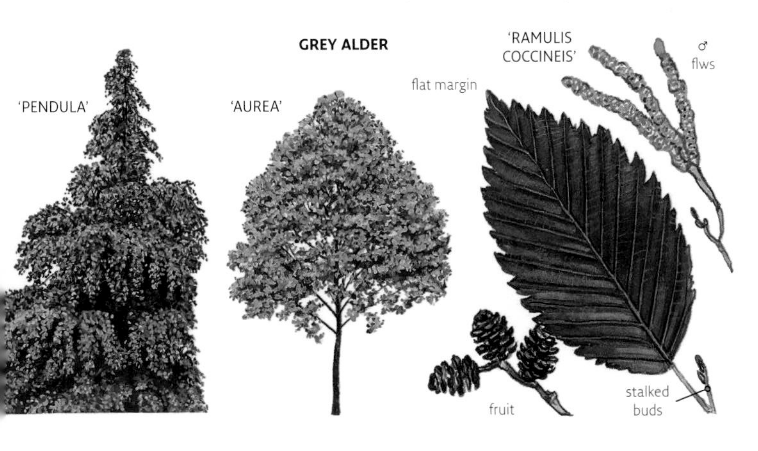

Male catkins to 10 cm, fawn-yellow and showy in spring. **Cones** Big (to 3 cm).

OTHER TREES Caucasian Alder, *A. subcordata* (1838), remains rare. Craggier bark; downy shoots; leaves more *oblong*, the base rounded/very slightly heart-shaped; *hairy* at least under the veins.

Spaeth's Alder, *A.* × *spaethii*, a hybrid (Berlin, 1908) of Caucasian with Japanese Alder (*A. japonica*), is rare but very handsome. Leaves *boat-shaped*, to 15 × 6 cm, dark, glossy; a few hairs beneath; small, *distant teeth* and *distant, curving vein*. Japanese Alder itself (in some collections) has smaller, *hairless*, long-pointed leaves, a gaunt habit, and bark cracking into large squares.

Oriental Alder, *A. orientalis* (Cyprus; the Middle East; 1924), is very rare. Bark grey-brown, soon harshly square-cracked; shoots hairless; oblong leaves smaller than Caucasian Alder's, rather dull above but usually *shiny beneath* and hairless except for tufts under the vein-joints.

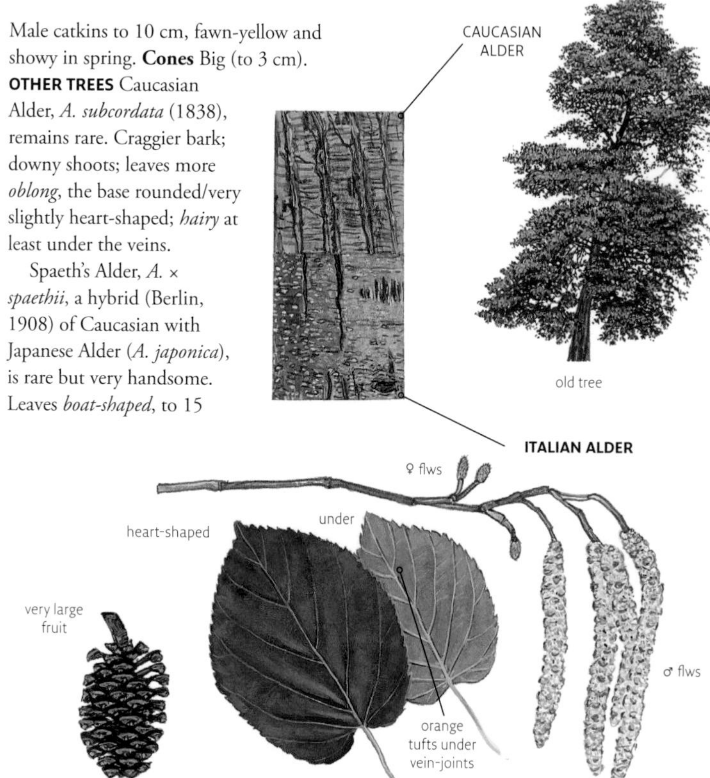

HORNBEAMS

Hornbeams (up to 70 rather similar species) hail mostly from East Asia. Their bark is usually smooth and grey; their leaves have many parallel veins. Catkins are hidden until spring; short strings of small nuts are backed by long, green, variously lobed bracts. (Family: Betulaceae.)

THINGS TO LOOK FOR: HORNBEAMS AND HOP-HORNBEAMS

- Buds: How long and how sharp? Appressed?
- Leaves: How many veins? Are they hairy beneath?
- Fruit: Nut-bracts – are they lobed (with what toothing? wrapped around the nut?)?

Common Hornbeam
Carpinus betulus

Europe including SE England; Asia Minor. In its native range often the dominant tree on heavy clay, having been selected and coppiced for charcoal production; locally abundant on richer soils in hedges and scrub and much planted throughout Britain and Ireland. Hornbeam means 'hard tree', and anybody who has tried sawing one will know why: the timber was used for chopping blocks, cog-wheels – and as an exceptionally calorific firewood.

'FASTIGIATA'

APPEARANCE
Shape Branches seldom heavy or spreading, though old trees in the open (to 30 m) can be very broad and shapely; delicate traceries of fine twigs. **Bark** Grey, smooth, with vertical, wriggling dull silver or orange snake-marks (cf. *horizontal* markings of young beech); odd wide fissures and shallow, smooth criss-crossing ridges develop with age. Trunks soon grow *lumpy, muscular flutings*. **Shoots** Slender, with long hairs at first. **Buds** *Slender* and long (8 mm), the tips *slightly incurved*; never spreading widely from the shoot like beech-buds. **Leaves** 7–12 cm; double-toothed, and corrugated by the 10–13 pairs of close, *impressed* veins; there are some long hairs above them, and on their stalks. Pale yellow in autumn. **Flowers** Male catkins expand in spring as yellow curtains. **Fruit** Nut-bracts 3 cm long, randomly toothed; *one short basal lobe on each side*.

COMPARE Common Beech (p.113): different foliage, but more alike in winter. Its three-lobed bracts *or* its long buds distinguish Common Hornbeam from most other hornbeam species.

VARIANTS 'Fastigiata' ('Pyramidalis'; 1885) is an abundant street tree, to 24 m, with a neat, dense, stylish *ace-of-spades shape* (but ultimately a broad balloon, and bushy if not trained to a 2 m stem in the nursery). Many light branches spring *from a single point* and curl in at their tips – the branches of woodland-grown wild trees can approximate to this habit, but will not all leave the trunk together.

'Columnaris' is rare, making a *densely rounded shape*, with big, broad, *very oblong* leaves.

COMMON HORNBEAM

European Hop-hornbeam
Ostrya carpinifolia

SE France E to the Caucasus. 1724? Very occasionally seen in larger gardens. **APPEARANCE** Often passed over as a Common Hornbeam (opposite); but conversely hornbeams in full fruit are sometimes assumed to be hop-hornbeams. To 24 m. **Bark** Brown-grey, cracking into *square, ultimately shaggy plates.* **Buds** Fat, *not pressed against the shoot* like Common Hornbeam's. **Leaves** Very like Common Hornbeam's but with a few more main veins (12–15 pairs); these *may branch.* **Flowers** Male catkins *exposed through winter* before they open (as in birches and alders). **Fruit** In *hop-like clusters of bladders* studding the crown almost like flowers through summer, *white*/greenish then red-brown; nuts (*sealed* in each bladder) 6 × 3 mm. **OTHER TREES** Ironwood (American Hop-hornbeam), *O. virginiana* (E USA, 1692), is in some big gardens: its leaves often have fewer veins and more long hairs, each with a tiny *swollen tip* (gland), and its nuts are larger (to 8 mm).

Japanese Hop-hornbeam, *O. japonica* (1888), is the rarest of the three in the UK; its leaves have fewer, distant veins (9–12 pairs), and are more *closely hairy underneath*.

EUROPEAN HOP-HORNBEAM

HAZELS

Hazels (about 15 species) expose their male catkins through winter; female flowers sit on the twigs like crimson sea-anemones and lead to large nuts in leafy cups. (Family: Betulaceae.)

Common Hazel

Corylus avellana

Europe including Britain and Ireland; Turkey; N Africa. Very abundant, except on poor or waterlogged soils, as a woodland understorey and in hedges. Once much planted for coppicing; the wand-like three- to six-year-old growths – so pliable that a strong hand can knot them – were indispensable for wattle hurdle-making and (in the south) to bind laid hedges. Hazel's fertility is now compromised in much of Britain by introduced Grey Squirrels, which habitually strip the nuts before they ripen.

APPEARANCE Shape Usually multi-stemmed; old plants to 15 m. **Bark** Often burnished bronze when young but harsh to the touch and finely peeling; old stems pale brown, with some shallow, flat ridges. **Shoots** Pale green-brown, with long, rather harsh hairs. **Buds** Green, *fat, oval.* **Leaves** Soft, hairy and floppy; nearly *round* (to 12 cm), with a *sudden sharp point*; on short (1 cm), long-hairy stalks. **Flowers** The yellow male catkins expand and open in late winter. **Fruit** Squirrels permitting, the nuts ripen early in autumn. They are sheathed in shucks *about their own length.*

COMPARE Filbert (below). Wych Elm (p.128): sometimes confused as bushy regrowth, but its thicker, rougher leaves have asymmetrical bases.

VARIANTS Weeping Hazel, 'Pendula', is rare: an umbrella of shoots from twisting limbs on a single, trained stem or grafted on a trunk of Filbert.

Corkscrew Hazel, 'Contorta', is occasional and much used in flower-arranging: its leaves and shoots curl madly (cf. Corkscrew Willow, p.90).

Golden Hazel, 'Aurea', is rather rare but vigorous (to 11 m): yellowish young foliage fades to dull pale green through summer.

Filbert

Corylus maxima

Balkans; Turkey. 1759. Occasional in old gardens.

APPEARANCE Shape A more vigorous plant than Common Hazel. **Bark** *Greyish and distantly cracked.* **Leaves** Often more distinctly lobed. **Fruit** The longer, narrower nuts are hidden in 'Christmas stockings' which are nearly *twice their length* and jaggedly toothed. Cob-nuts (cottage gardens; a few commercial plantations in W Kent) are mostly hybrids with Common Hazel.

VARIANTS Purple Filbert, 'Purpurea', is frequent and has purple leaves and catkins. (This is sometimes misnamed *Corylus avellana* 'Purpurea', a plant which does exist but is very rare and has brownish-pink *young* leaves and purplish catkins.)

PURPLE FILBERT

Turkish Hazel

Corylus colurna

SE Europe and Asia Minor. Grown in a few gardens since 1582 but now a locally frequent street tree.

APPEARANCE Shape A broad but symmetrical *spire. To 26 m*: a straight trunk which is rarely forked, carries light, level branches with dark, luxuriant, hanging foliage; broadening or leaning only in age. **Bark** Pale brown, with *close scaly ridges* from the first – rather like Field Maple's (p.205). **Leaves** Recognizably a hazel's, but more lobed and shinier. **Fruit** Like nuts of Common Hazel but a little larger; in frilly, bristly cups.

COMPARE Wild Service (p.163); Yunnan Crab (p.173).

TURKISH HAZEL

SOUTHERN BEECHES

The southern beeches are a genus of at least 40 southern-hemisphere trees, with beech-like nuts and diverse leaves. (Family: Nothofagaceae.)

THINGS TO LOOK FOR: SOUTHERN BEECHES

- Bark. What is it like?
- Leaves: Deciduous? How many main veins? Do they have teeth (regular? how big?)?

Antarctic Beech

Nothofagus antarctica

(Nirre) S Andes – to Cape Horn. 1830. Occasional.
APPEARANCE Shape *Sparse, gaunt, irregular*, like a lapsed bonsai, to 20 m; crisped peninsulas of fine twiggery; often bushy and usually leaning. **Bark** Dark brown, with lenticel-bands, then rough, grey plates. **Leaves** Tiny (2–4 cm), crinkled; *three to five pairs* of veins and large, blunt teeth *each with four small teeth*; dark, shiny, and looking evergreen, they flush early with a powerful scent of cinnamon and cheap soap.
COMPARE Red Beech: *simple* teeth.

Rauli

Nothofagus alpina

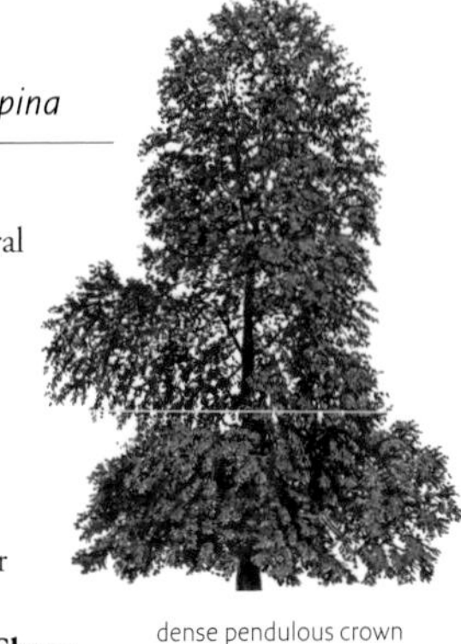
dense pendulous crown

(*N. procera, N. nervosa*) S central Andes. 1910. Occasional, doing best in mild, wet areas and freely naturali-zing. The fine timber is beech-like.
APPEARANCE Shape Usually strictly conic when growing fast; later with rather level

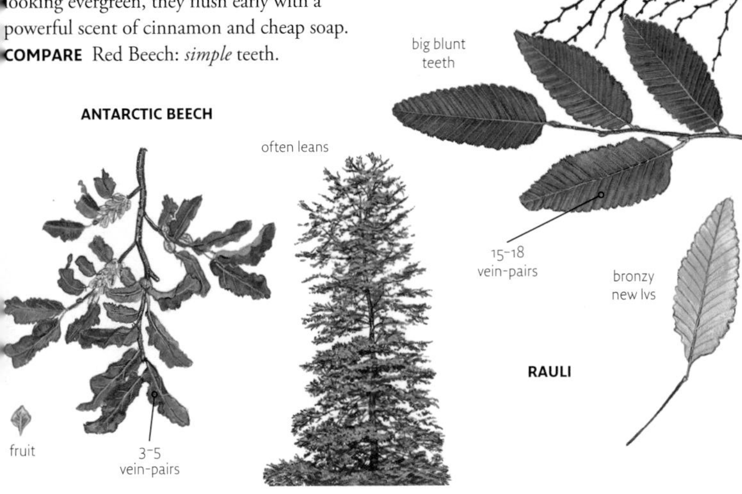

branches on a *strong often straight trunk*, and dense foliage. To 36 m so far. **Bark** Grey; soon with *long, flat, vertical plates*. **Shoots** Green, *long-hairy for a year*; sturdier than Roblé Beech's. **Buds** Birch-like: big (1 cm). **Leaves** To 9 cm; *15–18 pairs* of boldly impressed veins and *blunt* teeth/ little lobes. They flush bronze, and in autumn turn pale gold with some reds.
COMPARE Roblé Beech (below). Blunt toothing distinguishes from Hornbeam, etc. (see pp.107–8).

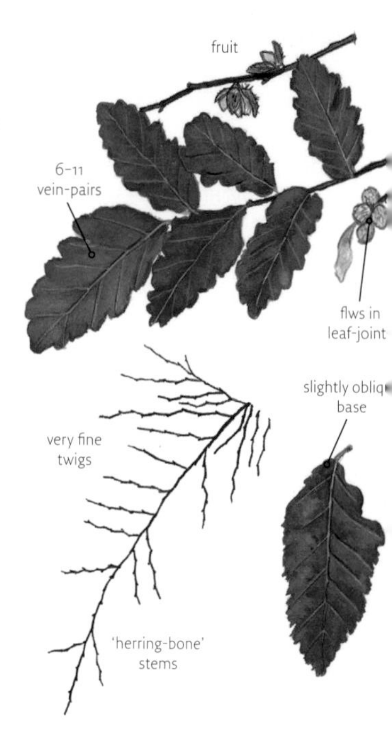

Roblé Beech
Nothofagus obliqua

(Coyan; Hualle) S Andes. 1902. Locally frequent (some plantations): a hugely vigorous (but short-lived) tree which seems to 'belong' in N European landscapes. Readily naturalizing.

APPEARANCE Shape Wispily conic, then openly *irregular*, to 30 m; seldom broad; fine, downward-fanning branches. **Bark** Brown/silvery, soon with square/rounded *harshly curling plates*. **Shoots** As slender as any tree's; fine white hairs. **Buds** Appressed, 4 mm – trees can look dead until these expand, branch by branch, in late winter. **Leaves** Small (4–8 cm); *6–11 pairs* of sunken veins; *sharply, irregularly toothed/* slightly lobed. Crimson and yellow in autumn.
COMPARE Rauli (above). The putative hybrid (*N.* × *dodecaphleps*; in odd collections) is intermediate. Irregular toothing distinguishes from all oaks/zelkovas.
OTHER TREES Hualo, *N. glauca*, is a lovely, tender Chilean tree in a few collections: bright *orange papery bark*; leaves (often heart-shaped at the base) *pale blue-grey* beneath.

BEECHES

Beeches (all northern-hemisphere trees) have smooth grey bark, long buds and edible nuts in four-parted husks. (Family: Fagaceae.)

Common Beech ✪
Fagus sylvatica

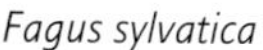

S and W Europe, including (probably) S and Middle England; planted abundantly throughout Britain and Ireland in plantations, parks and shelter-belts, and well naturalized; often dominant on mineral soils, but disliking wet ground. Sadly this is one tree whose younger bark frustrated male Grey Squirrels now habitually rip off in their breeding season, preventing tall straight growth.

APPEARANCE Shape In woodland often with a long, slightly sinuous trunk, but readily growing a huge dome on strong, smooth limbs; one of the UK's tallest broadleaves, to 40 m. The singularly dense foliage shades out competition and shields the thinly clad under-bark from the sun's heat. Many old trees are pollards, but beech is shorter-lived than most big trees, blowing down easily and quickly decaying. Seldom successfully coppiced. Twigs within 3 m of the base retain their dead leaves in winter (as do hornbeams and some oaks), so make a perfect hedge. **Bark** Silver-grey, with slight *horizontal* etchings; some trees develop shallow or rugged criss-crossing ridges. **Shoots** Slender, grey (silky at first) and zig-zag. **Buds** *Torpedo-shaped, 2 cm, copper/grey, spreading at 60°* (some may be a camouflaged snail, *Cochlodina laminata*). **Leaves** To 10 cm, with odd, tiny, distant teeth; hair-fringed, and silky all over as they unfold; 5–9 vein-pairs. **Fruit** Nuts in prickly husks, on 2 cm stalks.

COMPARE Hornbeam (p.107): often confused.

VARIANTS Many and popular. Copper Beech or Purple Beech, f. *purpurea*, is abundant and crops up frequently in the wild and in plantations, in colours ranging from a quiet pinkish brown (f. *cuprea*) to royal purple, depending on the proportion of purple xanthocyanins in the leaves. Ungrafted trees of dubious origin become dull and dark after the first flush of wine-red, with inner leaves dark green; named clones include the bright 'Rivers' Purple' (Sawbridgeworth; by 1870). In winter, a grafted beech more than 25 m tall is very likely to be Copper; a *single stem often persists*, with *numerous*, light, *ascending* branches from swollen bases and a crown *narrowing to a small flat top*. The nut-husks are also purplish.

Dawyck Beech, 'Dawyck', was discovered around 1860 in a wood next to Dawyck Gardens near Peebles by the head gardener, and since the 1930s has become a frequent municipal and garden tree. The branches rise vertically but twist and turn, like a Lombardy Poplar reflected in choppy water; mature trees (to 28 m) remain very narrow only in shelter. In winter the steel-grey cast and long buds distinguish it from Cypress Oak (p.120) and other big fastigiate trees.

Golden Beech, 'Zlatia' (1892, its name deriving from the Serbo-Croat for 'gold'), is undeservedly rare: the fresh yellow of the young leaves fades by mid-summer but is more even than the yellowing often seen on drought-stressed wild beeches.

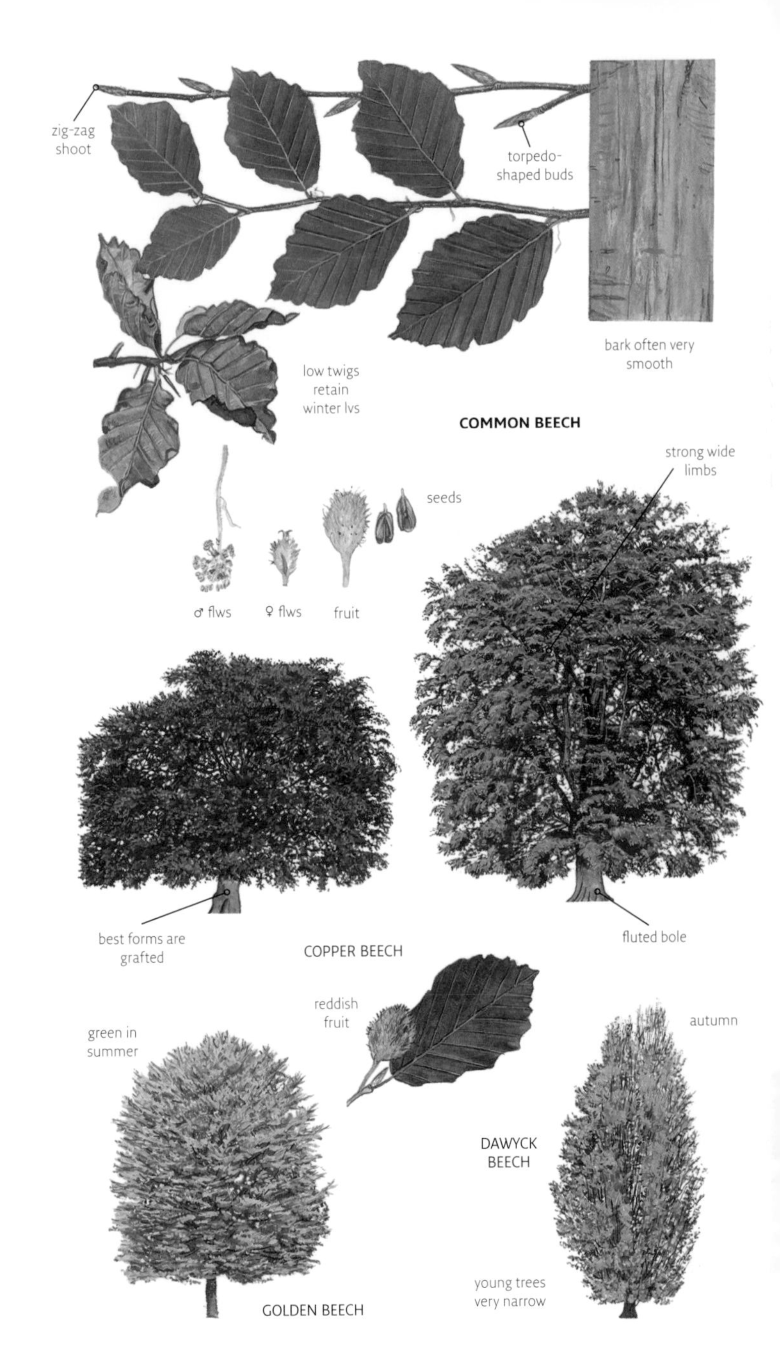
zig-zag shoot
torpedo-shaped buds
bark often very smooth
low twigs retain winter lvs
COMMON BEECH
strong wide limbs
seeds
♂ flws
♀ flws
fruit
best forms are grafted
COPPER BEECH
fluted bole
reddish fruit
green in summer
autumn
DAWYCK BEECH
young trees very narrow
GOLDEN BEECH

ern-leaved Beech, 'Aspleniifolia' Heterophylla'), is only locally frequent ut is a tree of great distinctiveness and eauty, to 28 m, generating interest and ometimes bewilderment. The depth of the bing varies from clone to clone. In the ommonest and most feathery form (seldom rafted), the shoot-tip leaves are narrower r even linear, and the crown is distinctively *ale, matt and fluffy* even when seen at distance; it colours early in autumn. 'his tree is a 'chimaera', with inner issues of typical Beech enveloped y cells of the sport, so that sprouts

with normal leaves will often grow from the trunk and branches, especially after an injury; unlike ordinary reversions, these seldom or never take over the whole crown. In winter, the tree is typically broad with a skirt of fine branches almost sweeping the ground, and has very *dense*, fine, horizontal or slightly rising shoot-systems; the distinctive leaves are also very slow to rot. Compare Common Alder 'Imperialis' (p.104); Cut-leaved Zelkova (p.138). No oak has leaves as small and as deeply lobed, or the characteristic 2 cm beech buds.

Weeping Beech, *f. pendula*, is the grandest of pendulous trees, to 25 m; locally frequent. Densely leafy shoots cascade from parabolically curving limbs; these layer given the chance, and the original tree brought from France to the Knaphill Nursery in Surrey in 1826 has grown into a whole cathedral of columns with walls of foliage 60 m apart. A few clones (producing many weeping seedlings) are more symmetrical, with shorter hanging shoots from widely arching branches.

SWEET CHESTNUT

Sweet chestnuts (10–12 species) have big nuts in spiny sheaths. Chestnut blight (Endothia parasitica), a fungal infection accidentally spread from East Asia (and still spreading northwards across Europe), has made the American Chestnut (Castanea dentata) almost extinct in its native range. (Family: Fagaceae.)

Sweet Chestnut
Castanea sativa

(Spanish Chestnut) S Europe, N Africa; Asia Minor; brought to Britain by the Romans and completely naturalized in warmer parts (though not elsewhere in N Europe). Abundant except on wet or very chalky soils, living to a huge age; dominant in many sandy woods in SE England where it was planted and coppiced for its timber: the vigorous, straight stems begin to grow a very durable heartwood after two to five years.
APPEARANCE **Shape** Often slender for a big tree, at least near the top, with a long bole in woodland, to 36 m; dense, glossy foliage. **Bark** In young trees dull silvery-purple, horizontally banded then vertically cracked. By 60 years brown, with a vertically stretched network of crisp, criss-crossing ridges which eventually *spiral* (usually clockwise). **Shoots** Grey, the tip aborted (cf. limes); knobbly – on stronger growths – from *long prominent buttresses* below each blunt, few-scaled, hairless bud. **Leaves** Bigger than any other wild tree's, with *spine-teeth 1 cm apart*; scurfy underneath, at least at first. **Flowers** Crowns turn fawn in high summer with stiffly spraying male catkins which smell strongly of frying mushrooms; good crops of chestnuts follow only in warm parts. (Fruiting chestnuts – 'Gros Merle', 'Paragon', 'Marron de Lyon' – are grafted trees with usually only one nut per husk; very scarce in the UK.)
VARIANTS Variegated Chestnut, 'Albomarginata' (1864), is a rare but bright, delicate tree, apt to revert. The creamy leaf-margin is yellow as the leaves unfold.

'Variegata' ('Aureomarginata'), whose leaf-margins remain gold, is much rarer to date.

Cut-leaved Chestnut, f. *heterophylla*, is a group of rare, often feeble clones whose strongly toothed leaves can (especially at branch-tips) be *reduced to strips*. 'Laciniata' (1838) is distinct, if liable to revert: all its teeth are *drawn out into long threads*.
OTHER TREES American Chestnut, *C. dentata*, is confined in the UK to a few collections. Its leaves are slightly narrower and more tapered at the base than Sweet Chestnut's, and *completely hairless*, but its buds are finely downy.

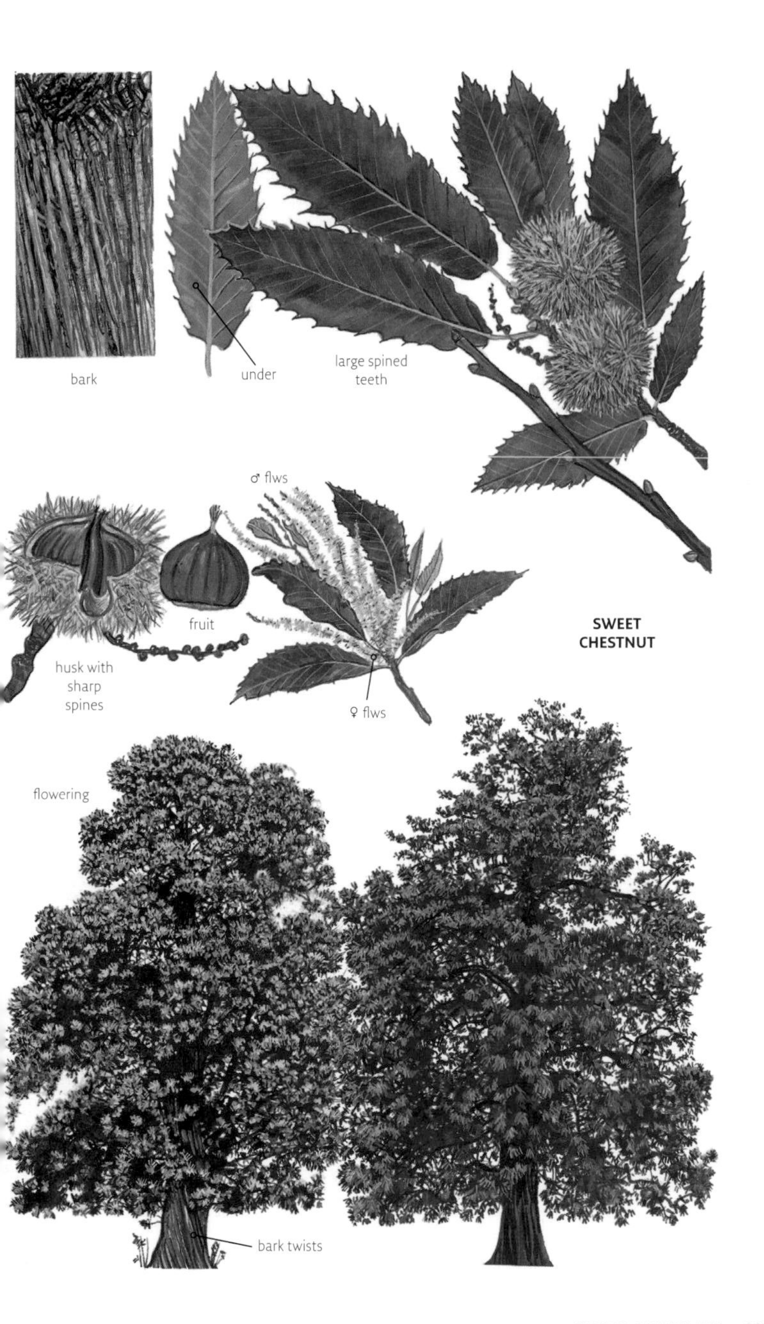
bark
under
large spined
teeth
♂ flws
fruit
SWEET
CHESTNUT
husk with
sharp
spines
♀ flws
flowering
bark twists

OAKS

Oaks (500 species) all carry 'acorns' in 'egg-cups'. Their buds cluster at shoot-tips (any can dominate next season to create the twisting, wide limbs of many species). (Family: Fagaceae.)

Things to Look for: Oaks

- Shoots: Hairy?
- Buds: Whiskered?
- Leaves: Evergreen? Shape (especially at base)? Any bristle-tipped teeth? Hairs/wool underneath? Leaf-stalk – how long is it?

Key Species 

Sessile Oak (below): rounded, untoothed lobes. **English Oak** (opposite): similar leaves with auricles. **Turkey Oak** (p.120): jaggedly lobed leaves (fine felt underneath). **Red Oak** (p.125): lobes each with several whiskered teeth. **Holm Oak** (p.123): evergreen leaves, untoothed at maturity. **Lucombe Oak** (p.122): evergreen, lobed leaves.

Sessile Oak

Quercus petraea

(Durmast Oak) Mid Europe, including Britain and Ireland. Much more locally dominant than the familiar English Oak (opposite), avoiding heavy/alkaline soils, and absent from many lowlands. Seldom planted in forestry or until recently for ornament, for all its poise and stature.
APPEARANCE **Shape** Cleaner and less twiggy than English Oak; larger, glossier leaves *evenly spread.* Often taller (to 42 m), but equally gigantic with age. **Bark** As English Oak; can be more shallowly scaly. **Shoots** As English Oak. **Buds** Tend to have more scales. **Leaves** With regular, rather shallow lobes; main veins hairy beneath at

first and generally only running to the lobe-tips. Base *broadly tapered* (only sometimes and faintly showing the backward-pointing auricles of English Oak), on a *12–20 mm stalk.* **Fruit** Acorns sit on the twigs, *with short stalks or none* ('sessile').
COMPARE English Oak (opposite). Downy Oak (p.121): leaves much hairier beneath.

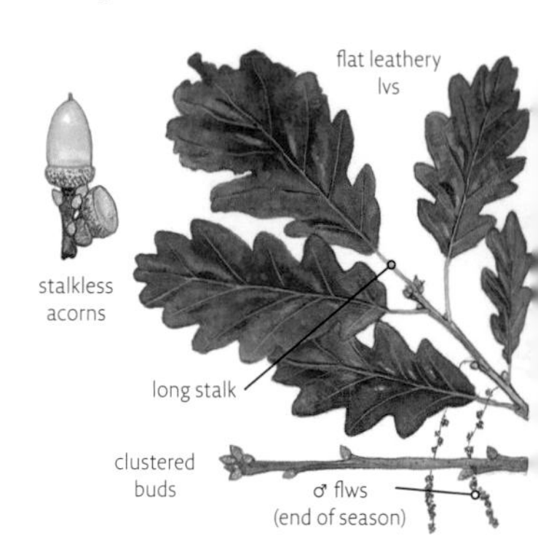

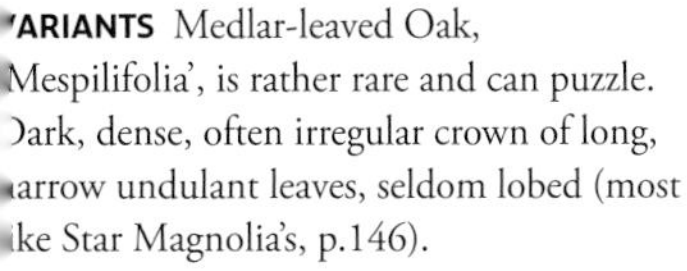

ARIANTS Medlar-leaved Oak, Mespilifolia', is rather rare and can puzzle. ark, dense, often irregular crown of long, arrow undulant leaves, seldom lobed (most ke Star Magnolia's, p.146).

nglish Oak ✪
uercus robur

Pedunculate Oak; Common Oak; *Q. edunculata*) Europe, including Britain and reland, to the Caucasus. Abundant except n marshy, chalky or very light soils; the dominant big tree across much of Britain, supporting a greater variety of leaf-eating insects than any other. Planted and/or pollarded (along with Sessile Oak) in old deer parks. Often the most valuable of forestry trees, with selected 'standards' mixed through most coppiced woods; everywhere in parks and gardens. Lifespan 1000+ years. In contrast to Sessile Oak a 'weed' species in scrub etc., though currently reluctant to reproduce within woodland: seedlings in shade succumb to the mildew (*Microsphaera alphitoides*) which can turn mature crowns grey late in hot summers.

APPEARANCE Shape Heavy, spreading, *twisting* branches make a broad crown, to 38 m; foliage in *blobby bunches.* **Bark** Grey; short, deep, knobbly ridges. **Shoots** Silvery; orange-brown oval buds cluster at their tips. **Leaves** With *irregular, deep lobes* (pointed only on sprouts); *two tiny lobes at the base* (auricles) flank the *short* (4–10 mm) stalk. Small veins extend towards the *deep, narrow sinuses* between each lobe. **Flowers** Curtains of yellow male catkins as the leaves unfold orange. **Fruit** Acorns (skipped in many years in N Europe to

limit populations of acorn-predators) often paired, on a *5–12 cm stalk* ('peduncle').

COMPARE Sessile Oak (p.118). The hybrid (*Q.* × *rosacea*) is probably occasional (in continental Europe, Sessile Oaks tend to flower a fortnight after English Oaks, making crosses rare). Turkey Oak (below): even as a seedling, has long-whiskered buds. Among widely grown oaks, only English Oak have pronounced 'auricles', but exotic species raised from seed may turn out to be hybrids with English Oak, with auricles but usually bigger leaves, less deeply lobed.

VARIANTS Cypress Oak, f. *fastigiata*, is at its best often taken in winter for a Lombardy Poplar (p.81), but the dense shoots are thicker and the branches twist; solid and blackish in leaf – like an Italian Cypress (p.34). Many are broader and open, but with *erect writhing shoots from steep branches*. Occasional; to 30 m.

Cut-leaved Oak, 'Filicifolia', is rare: a feathery, ghostly-grey tree, often gaunt, to 17 m. Hairs under its young leaves suggest it is a clone of *Q.* × *rosacea*.

Purple Oak, 'Atropurpurea', is rare and stunted. Trees with leaves flushing red-purpl then green (f. *purpurascens*) grow larger and may be seen in the wild. They redden again during the species' flush of 'lammas-growth' in high summer (which repairs the ravages o spring leaf-eating caterpillars).

Golden Oak, 'Concordia' (1843), has yellow foliage, fading through summer; ver rare. (Yellow chlorotic foliage is often seen on oaks, but all the growths will never be s evenly coloured.)

Turkey Oak ✪

Quercus cerris

SE France E to Turkey. 1735. Abundant in parks and belts; often aggressively colonizing on sandier soils. Host to alternate generation of the Knopper Gall wasp (*Andricus quercuscalicis*), whose caterpillars now turn the acorns of English Oaks within flying-distance into oozing, lumpy galls: eradicatio of Turkey Oak from the New Forest is consequently planned. Vigorous and very tough,

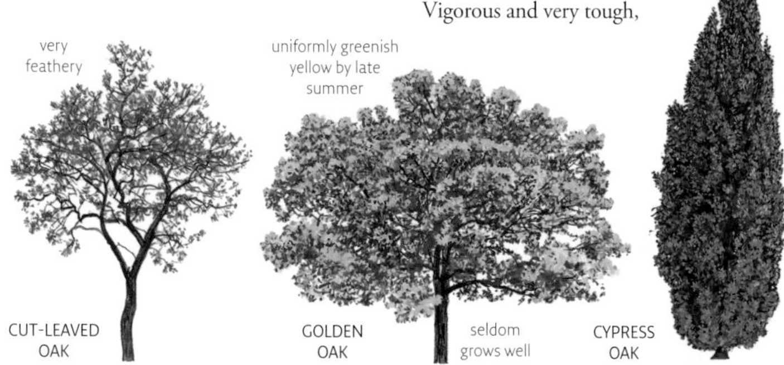

ıriving in coastal exposure, but
orthless as a timber tree.
PPEARANCE Shape
ometimes with a long,
raight bole and pointed
p; to 40 m. Straighter,
enderer branches than
ild oaks' are swollen
here they leave the
unk. *Blackish* and
ightly feathery in summer;
ıe clustered, mop-head buds tipping
ıe straight shoots are conspicuous in
inter. **Bark** Palish mauve-grey: *deep,*
edge-shaped fissures, often orange at the
ase. **Shoots** Slender; densely grey-hairy.
uds *All* with *big twisting whiskers*. **Leaves**
hick; rough but quite shiny above and
ninutely grey-felted beneath. Slender,
ut variable: often with simple, pointed/
ounded lobes but sometimes elaborately cut
strong growths/f. *laciniata*). **Fruit** Acorns
alkless, the cup-scales carrying mop-heads
f more *big whiskers*.
OMPARE Lucombe Oak (p.122): can
rade towards its parent but usually has
ınwhiskered side-buds, regular triangular
obes to *evergreen* leaves, and shorter,
wisting branches.

Downy Oak
Quercus pubescens

(Green Oak; *Q. lanuginosa*) S Europe, W Asia; often dominant, especially on limestone. Long grown in the UK but now rare.

APPEARANCE Shape Wide, twisting branches like English Oak's (p.119); to 22 m. **Bark** Almost like Holm Oak's (p.123): black-grey and quite closely square-cracked. **Shoots, Buds** Softly grey-downy.

Leaves Long stalks (like Sessile Oak's (p.118) but hairy), and the irregular lobes and deep narrow sinuses typical of English Oak: they are downy above at first and normally remain densely woolly beneath. (Some in cultivation may be hybrids – their leaves are almost hairless beneath by autumn.) **Fruit** Acorns nearly stalkless; cup-scales grey-downy.

Lucombe Oak

Quercus × hispanica 'Lucombeana'

(*Q.* × *crenata* 'Lucombeana'; *Q. lucombeana*) Turkey and Cork Oaks hybridize freely in S France/Italy, but the locally frequent trees in Britain and Ireland mostly derive from a cross raised at his Exeter nursery by William Lucombe in 1762. This is fertile, and today's population incorporates backcrosses with both parents; seedlings can also be found. The nomenclature is confused, but several variants can be distinguished. The original clone (provisionally 'William Lucombe') is the most frequent, especially in E Devon.

APPEARANCE Shape Tall (to 35 m): *heavy, twisting branches* are much swollen as they leave the trunk, which is *never long*; interior clean and *open*, surfaced with shining dark leaves which look deciduous but *hang on thinly until spring*. **Bark** *Purple*-grey, with deep triangular ridges; not corky. **Shoots** Hairy-grey; the end bud has big whiskers, the side-buds *never do*. **Leaves** Finely grey-felted beneath; the rather regular triangular lobes have *minute whisker-tips*.

COMPARE Turkey Oak (p.120); differences emphasized).

VARIANTS 'Crispa' (1792) is almost as frequent; it differs from Cork Oak (p.124) in stronger growth and more substantially lobed leaves. A broad, low and dense tree, with hanging outer shoots and many wandering twisting branches; fully evergreen. Bark cream/grey; soon *deeply corky*. Leaves small (7 cm); oval, with blunter lobes than 'William Lucombe'; a small proportion reduced to straps.

Fulham Oak 'Fulhamensis' (distributed by Whitley and Osborne of Fulham by 178 and perhaps of independent origin; now certainly rare) resembles 'William Lucombe' but has a corkier bark, weeping outer shoots and neatly oval leaves with usually six pairs of *precise, triangular lobes*.

'Cana Major' (1849) is a fully deciduous clone.

Holm Oak ✪ 💧

Quercus ilex

(Mediterranean Oak) S Europe; introduced early and abundant away from the coldest parts in shrubberies and belts, especially near the sea (where it copes outstandingly with salt spray); quite well naturalized in milder areas.

APPEARANCE Shape Often bushy; closely rising, rather *straight* branches from a sinuous bole; to 30 m. Very dense; the outer shoots may weep. **Bark** Blackish; becoming closely and shallowly square-cracked. **Shoots** Slender, with fawn wool. **Buds** Tiny; the end one has curling whiskers. **Leaves** Evergreen: blackish, a *fawn-grey felt* coating the often concave underside; spinily lobed on sprouts and saplings, then untoothed. Very variable in breadth. **Flowers** Fireworks of golden male catkins in early summer as last year's leaves drop. **Fruit** Acorns small (15–20 mm), with felted cup-scales. The acorns of var. *ballota* (*Q. rotundifolia*),

whose leaves are smaller and more rounded, are *edible*: once much grown in S Spain and N Africa.
COMPARE Cork Oak (below). Phillyrea (p.241): similar bark and crown but leaves, *in opposite pairs*, hairless. Most evergreen oaks have longer, elegant, glossy leaves.

Cork Oak

Quercus suber

CORK OAK

Mediterranean Europe from Portugal to Croatia; Morocco. Long grown in the UK: very occasional, though hardy and much admired. In Mediterranean plantations the cork is stripped off every seven to ten years without damaging the underlying cambium, though the advent of plastic corks has threatened this ancient, wildlife-rich landscape.
APPEARANCE **Shape** A low, dark, matt dome on heavy, *twisting* branches; to 23 m. **Bark** Cream, orange or grey, soon *deeply convoluted* like a Brancusi sculpture. Annual rings can be counted in the broken edges. **Shoots** Woolly. **Leaves** Evergreen, with small, often spiny lobes; blackish above; *pale grey* beneath with a dense felt.
COMPARE Holm Oak (p.123): very different bark; leaves untoothed at maturity. Lucombe Oak (p.122): sometimes as corky; has larger, consistently triangular leaf-lobes.

Hungarian Oak

Quercus frainetto

(*Q. conferta*) The Balkans, Romania, Hungary and S Italy; the specific name is a corruption of the Italian word for the tree, 'farnetto'. 1837. Occasional, but beginning to be more widely grown. In many ways the grandest of all oaks.
APPEARANCE **Shape** Typically a broad and splendid glossy-leaved wheel of *straight,* rather stiff, radiating branches, to 38 m, creating a slightly jagged outline;

HUNGARIAN OAK

COMPARE Sessile Oak (p.118): the bark is sometimes finely cracked, but trees are more rugged and twisting (with *smoothly tiled* bud-scales).

oodland-grown trees can have long boles ıd are slenderer. **Bark** Pale/purplish grey, :acking into rather small, neat square lates. Many older trees (the clone now ıarketed as 'Hungarian Crown') are grafts n English Oak. **Shoots** Softly hairy, at least : first. **Buds** Large (1 cm), grey-brown, 'ith numerous loose, hairy scales, but no 'hiskers. **Leaves** Big (to 25 cm), *elaborately ıt into many rather square, narrow lobes*; ɔwny beneath and with a few harsh hairs bove; in autumn gold and biscuit-brown. Some trees are hybrids with English Oak: maller leaves with fewer lobes and slight uricles. Acorns collected in cultivation also ield hybrids with Turkey Oak.)

Red Oak ✪
Quercus rubra

(*Q. borealis*; *Q. maxima*) E North America. 1724. Abundant in warmer parts: parks, gardens, forestry rides (and in a few plantations). Growing as large in Britain as in its native habitat (like several other of the 'red oaks' but unlike the 'white oaks'); rarely naturalizing.

APPEARANCE Shape Soon broad, on *strong, wide but clean and straight branches*; to 32 m. The American 'red' and 'willow oaks' (collectively 'black oaks') lack the rugged, crooked appearance of many oak species; their timber is soft and they are relatively short-lived. **Bark** Silver-grey and smooth at first. The British population is now diverse and some old trees retain smooth bark, with hemispherical warts; others develop shallow or even scaly grey ridges between orange fissures. **Shoots** Slender, grey, quickly hairless. **Buds** Chestnut, the scales with slightly hairy tips. **Leaves** *Big* (often 20 cm long), the variably *shallow* lobes each with *two or more whiskered teeth*; soon *hairless*

RED OAK

fast-growing but short-lived

branches radiate

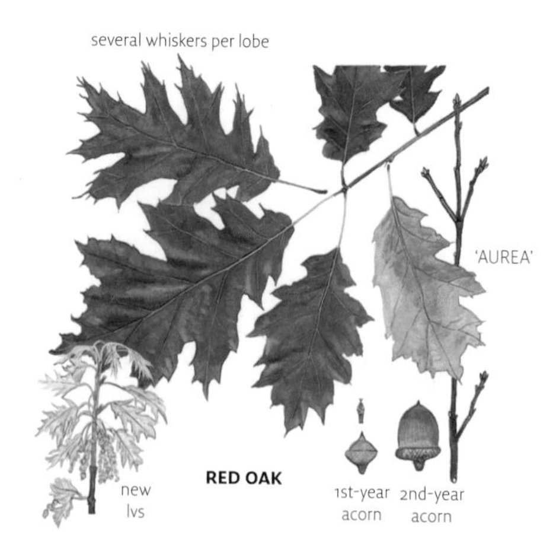

except for minute buff tufts under the vein-joints. Seldom glossy above and always a *matt pale green beneath.* They unfold late and are pale yellow for a week; autumn colours can be orange-brown or deep red, but are a disappointing warm brown in many seasons in the UK. **Fruit** Acorns, like those of other 'black oaks', are small (2 cm) and take two years to ripen.

COMPARE Other 'red oaks': Scarlet and Pin Oaks (below and opposite). These are slenderer trees with usually smaller, more deeply lobed leaves *glossy beneath.*

VARIANTS 'Aurea' has leaves which *remain* brilliant gold through early summer. Very rare.

Scarlet Oak
Quercus coccinea

SE and central USA. 1691. Rather occasional; rare in colder areas.

APPEARANCE Shape Irregular and quite slender, generally on a long, *sinuous* trunk with rather *few, long but slender wandering limbs*; untidy with ascending small branches and twigs. To 30 m. Sprinkled in high summer with *very yellow* late growths. **Bark** Silver-grey; sometimes remaining smooth except for hemispherical warts; more often is shallowly rugged and purplish, with orange fissures. **Shoots** Slender, soon hairless. **Bud** 5 mm, red-brown, the scales with hairy tips. **Leaves** Typically 13 cm long and *risin on each side of the stalk; deep, rounded bays* between perpendicular whisker-toothed lobes; glossy both sides, with *small* buff tuf under main vein-joints. Autumn colours smouldering reds at least near the branch-tip the most reliably colouring oak in the UK.

COMPARE Pin Oak (opposite): often most easily distinguished by crown shape. Its leaves have *big* tufts under the vein-joints; 'Splendens' (opposite) confuses the picture

but should always have a visible graft. Red Oak (p.125): bigger leaves more or less *matt* underneath. Black Oak: blackish, leathery leaves, often scurfy beneath.
VARIANTS 'Splendens' was selected for vigour and autumn colour at the Knap Hill Nursery, ırrey, by 1890 and has been much planted least in SE England. A grafted tree with gger leaves (to 18 cm) which have *bigger fts* under the vein-joints (cf. Pin Oak).

in Oak
uercus palustris

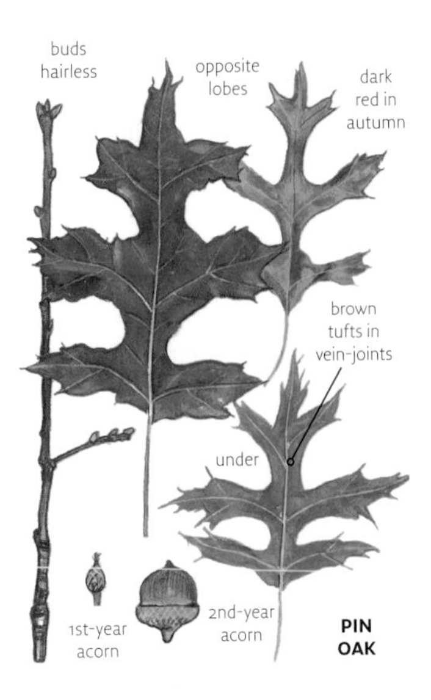

ntario to N Carolina and Kansas, in wet :es. 1800. Frequent in warm areas of the K, especially as a young tree, but almost ›sent from Scotland.
PPEARANCE Shape Usually very ıstinctive: trunk rather straight; crown :oad-conic then (in the open) densely omed, with many very fine dead branches tained in the interior and fine live ones forming a *distinct descending skirt from 5 m.* Old trees become more open and irregular after breakages, but often retain traces of the skirt, as do crowded woodland trees. To 28 m; the small leaves make the tree look taller. **Bark** Silver-grey; darker and more ruggedly ridged only in age. **Shoots** Slender, soon hairless. **Buds** Small (3 mm), *dull* brown, more or less *hairless*. **Leaves** Typically only 11 cm long, with deep, narrow, perpendicular lobes; fresh green, and glossy at least underneath; there are always *big* (2–4 mm wide) *buff drifts of hair* under the main vein-joints. Autumn colour a rich, uniform scarlet-brown in good years. **Fruit** Acorns in very *shallow* cups.
COMPARE Scarlet Oak (above): *small* tufts under the vein-joints, except in 'Splendens', and a more open, irregular crown. Red Oak (p.125): bigger leaves matt underneath.
OTHER TREES Northern Pin Oak, *Q. ellipsoidalis* (Ontario to Missouri, 1902), is in some gardens as a young, autumn-colouring tree; the cups of the *almost stalkless* acorns are *deeper* and less saucer-like than Pin Oak's, enclosing at least a third of the fruit.

ELMS

There are up to 60 species of elm. Their seeds are surrounded by a round, aerodynamic wing. All the forms included here, except for 'Ulmus × diversifolia' (p.130), have asymmetrical leaves, meeting the stalk consistently higher on one side than the other. Some forms have leaves scrubby above, like sandpaper, from very short stiff hairs; in the majority only juvenile leaves (on sprouts and low branches of older trees) are rough. Winter shoots are generally dark grey with darker, purple-brown buds. Since 1966, a new virulent strain of Dutch Elm Disease (DED) has destroyed most old elms in many parts of Europe. DED is due to a fungus (Ophiostoma novo-ulmi) that is transported from tree to tree on the mouthparts of bark-boring beetles (Scolytus species); in an attempt to isolate the infection, the tree shuts down its sap-conducting vessels, and the crown above the blockage is starved of sap and dies within days. The root system usually remains alive and most elms are able to sucker vigorously, but the new plants become vulnerable to infection after about ten years when their trunks are thick enough for beetle attack. (Family: Ulmaceae.)

Things to Look for: Elms

- Leaves (adult): How rough? How many secondary teeth? How downy beneath? How asymmetrical at the base? Stalk – how long?
- Fruit: Downy-winged? What shape? Stalk – how long?

Key Species 

Wych Elm (below): big, always scrubby leaves. **Huntingdon Elm** (p.133): big, glossy adult leaves. **Smooth-leaved Elm** (p.132): narrow, glossy adult leaves. **English Elm** (p.130): nearly round, variably scrubby leaves.

Wych Elm ✪
Ulmus glabra

(Scots Elm; *U. montana*) Europe; W Asia; the one elm indisputably native to Britain and Ireland. Abundant in upland areas; more local to S but much planted. Old trees now rare except in N/W Scotland: seldom suckering, so much reduced by DED.
APPEARANCE Shape Young plants broad, on sinuous stems. Old trees domed and billowing; thicker shoots than other wild elms (*never corky-winged*), and dull, black-green foliage. A giant tree (to 40 m) – even

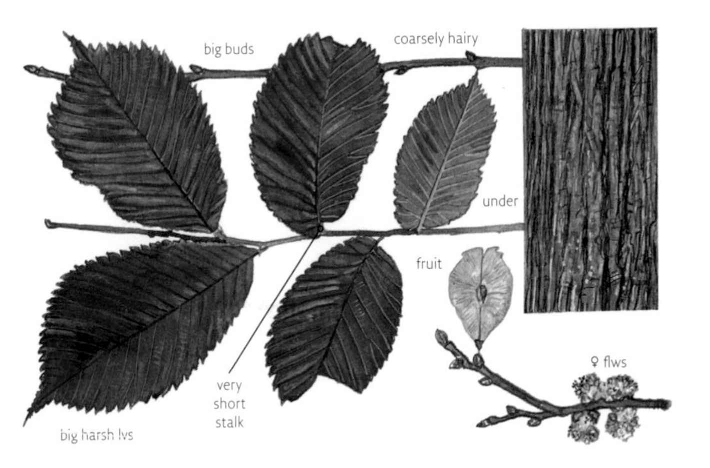

n coastal exposure. **Bark** *Smooth and grey for 20 years*; then grey-brown shaggy ridges.)ften sprouty. **Shoots** Dark grey, with hard ıairs. **Buds** Black-purple, hairy; *broad* and quat. **Leaves** The largest of any native tree's to 18 cm; 14–20 vein-pairs), hard, rather *blong* and often (though *never* in the N 3ritish/Scandinavian ssp. *montana*) with ıorn-like lobes at the 'shoulders': matt and *ermanently scrubby* above; some stiff, thin, vhite down underneath. *Almost stalkless* 2 mm on the 'short' side). **Fruit** Usually ıbundant: the wing downy at its notched ip, the seed central.

:OMPARE European White Elm (p.135).

/ARIANTS 'Camperdown' (1850) ometimes resists DED and is now a very occasional garden tree. A green igloo: branches writhe from a high graft, the hanging shoots dense with big leaves almost hairless beneath.

'Pendula' ('Horizontalis'; 1816) is also now very occasional. To 18 m, grafted; the finer branches in wide, *gently descending sprays*: stiff yet elegant like the fingers of a Javanese dancer.

Golden Wych Elm, 'Lutescens', is now rare. Leaves emerge soft spring-green and *intensify to brilliant yellow* through summer. (Much larger-leaved and more elegantly domed than 'Louis van Houtte', p.131, and other golden elms.)

Exeter Elm, 'Exoniensis' (by 1826), is now rare: a quirky balloon-shape (to 20 m)

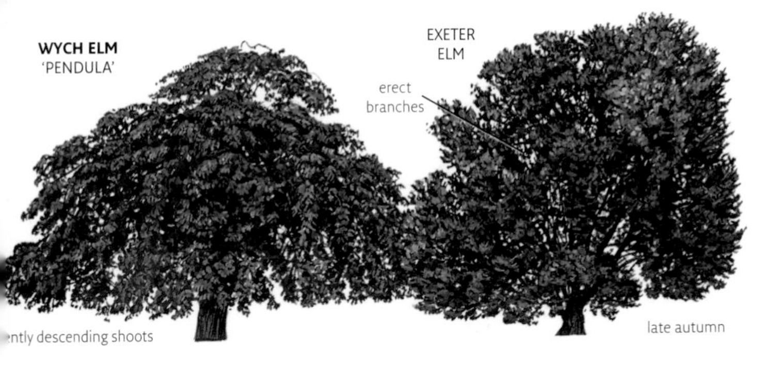

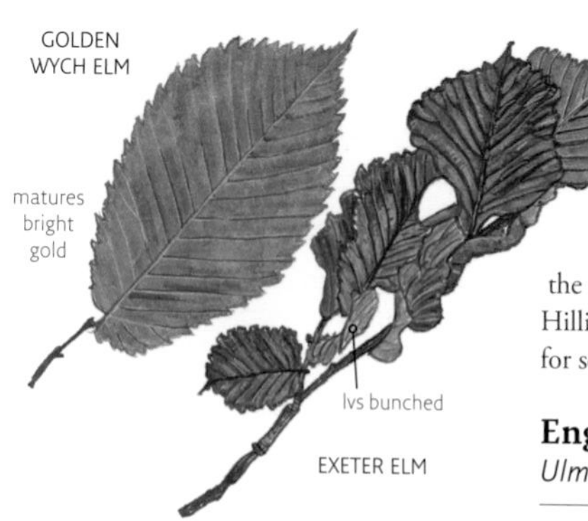

of twisting, *erect* branches, carrying dense bunches of smaller, rounded, twisted leaves (cf. 'Dampieri', p.133).

OTHER TREES Field Elm (*U. minor*; Europe; W Asia; N Africa) is the other species found wild in Britain and Ireland. It has (much) smaller/slenderer leaves than Wych Elm's, distinctly stalked and typically smooth except on sprouts; craggier bark, and hairless fruits, the seed near the end notch: the common elm of lowland England (invading some woods but scarcely a woodland tree), perhaps introduced as a fodder crop by early farming tribes. It spreads mainly by suckering so many local clones exist, several (e.g. 'Boxworth Elm' and 'Dengie Elm') showing considerable immunity to DED; only the most widespread and stable are treated on pp.130–3. Trees turn gold at the end of autumn and (except for English Elm) flush *very late*.

Hybrid Elms (*U.* × *hollandica*) occur when fertile Field Elms grow near Wych Elms; suckering clones dominate in parts of East Anglia, combining the parents' features variously (i.e. large, smooth adult leaves or smaller, scrubby ones, usually on distinct stalks), and tending to have some DED-resistance. Named types include '*U.* × *diversifolia*' (mainly Hertfordshire, Cambridgeshire and Suffolk), with *small*, very slightly rough leaves, downy beneath, and *all symmetrical at the base* on at least one side-shoot in ten, a rather square-cracked bark and an open, spreading crown; and '*U.* × *elegantissima*' from the E Midlands (a variable grouping to which the popular tiny-leaved bush 'Jacqueline Hillier' is believed to belong). See pp.133–5 for some ornamental clones.

English Elm ✪

Ulmus minor var. vulgaris

(*U. procera*; *U. campestris*). The common elm across the richer farmed soils of middle England (and the only Field Elm – see left – with often *permanently scrubby leaves*); once dominant in many landscapes, spreading by suckers but hardly ever by seed: the genetically identical trees turned out to be the most susceptible elms of all to virulent DED and have now been entirely reduced in the south (outside disease control zones) to locally abundant regrowth surviving as a rare planted tree further north and west.

PPEARANCE Shape Suckers spikily conic: egular fish-bone patterns of fine, stiff hoots; twigs in sun develop *corky wings* nuch more often than other Field Elms cf. Dutch Elm, p.134; Field Maple, p.205; weet Gum, p.151). Old trees (once to 5 m) tall-domed like thunderheads, on n often straight bole with many sprouts nd *a few big limbs*; blackish with a smoky nultitude of dense leaves. **Bark** Soon with lose, knobbly, pale brown ridges, then grey, *oarsely square-cracked*. **Shoots** Fine, downy. **uds** *Tiny*, grey/purple. **Leaves** *Often tay scrubby above*; finely downy beneath; *ather round* (cf. Coritanian Elm, p.133), –10 cm, and puckered/*crumpled*; on short 5 mm) stalks. The vein-hairs can sting, like he related Nettle's.

ENGLISH ELM
'LOUIS VAN HOUTTE'

VARIANTS 'Louis Van Houtte' (1880) is a rather spikily erect golden-leaved sport (cf. 'Dampieri Aurea', p.133); now almost extinct in gardens.

Cornish Elm

Ulmus minor var. cornubiensis

(*U. stricta*) A group of Field Elms locally abundant in Cornwall, W Devon and SW Ireland; nearly extinct as planted trees elsewhere. Typically narrow-domed on a straight bole; the branches rise quite steeply with (in exposure) sky visible between dense systems of *vivid green* foliage. The bark grows very scaly grey-brown ridges, which may *curl free at each end*. The shoots are finely hairy only at first; the leaves small (about 6 cm) and sometimes *cupped*; leathery-smooth above (but scrubby on suckers and low shoots) and downy only under the midrib; on good 1 cm stalks downy above. The teeth (with zero to two secondary teeth) are rather *blunt*.

CORNISH ELM

LOCK ELM

WHEATLEY ELM

Smooth-leaved Elm ✪
Ulmus minor var. minor

(*U. carpinifolia*) This name covers the common elms in parts of Suffolk, Essex, Hertfordshire, Kent and E Sussex (and across much of Europe, W Asia and N Africa; planted elsewhere); often spreading from seed, these are more varied than othe[r] Field Elms, and may resist DED.

APPEARANCE Shape Domed, on usually sinuous boles; dark, glinting foliage. Some forms weep dramatically (f. *pendula* – still widespread in Hertfordshire); some are stiff and straight. **Bark** Grey-brown; scaly sometimes criss-cross ridges develop slowly.

OTHER TREES Goodyer's Elm, var. *angustifolia*, from near the coast in Hampshire (and in Brittany) differs in its rounded, darker crown and longer-stalked leaves which have *two to three secondary teeth*; no mature trees are now known.

Wheatley Elm, 'Sarniensis' (Jersey Elm; Guernsey Elm), from the Channel Islands, differs from Cornish Elm in its blackish-green *spear-shape* (to 37 m) on light, steep branches (cf. 'Lobel', p.134), the tip narrowly rounded even on forking trees; its leaves have *one to three secondary teeth*. An abundant street tree since 1836; survivors are now rare.

Lock Elm, var. *lockii* (Plot Elm; var. *plotii*), from the N Midlands, is a slender tree unlike the Cornish Elm in its light, more spreading branches, *open, tilted tip* and weeping shoots: many *side ones*, instead of producing consistent rounded patterns of three to six leaves, *continue to grow* in plumes (cf. Asiatic elms, p.136–7). Its leaves are dull above and *minutely roughened*, and *white-hairy beneath* at least in patches near the midrib. Rarely planted outside the natural range and now very scarce.

hoots Soon hairless, occasionally corky-
inged (cf. English Elm, p.130). **Buds**
lender, downy. **Adult leaves** Glossy, flat
nd leathery-smooth; 6–15 cm (depending
n the clone); variably *narrow* and *narrowly
apering to the base on the short side*, on a
ood 1 cm stalk. Compare the new hybrid
lms (p.134–5).

ARIANTS 'Viminalis' is a rare slender wild
ariant which has *jagged, hooked 1 cm teeth.*

Coritanian Elm (rarely given specific
tatus – *U. coritana* – but often subsumed
n *U. minor* var. *minor*/var. *vulgaris*) has
cattered populations in E England (little
lanted elsewhere). Raggedly spreading; leaves
ounded (like English Elm's – but *always
mooth* except on sprouts and low branches; cf.
Dutch Elm, p.134) and often slightly *heart-
haped at the base*, on 1 cm slender curved
talks; up to four blunt secondary teeth.

Huntingdon Elm ✪

Ulmus × hollandica 'Vegeta'

(Chichester Elm) Raised in a Huntingdon
nursery *c.*1760. For long the most planted
rnamental Hybrid Elm (see p.130): it has
ome DED-resistance so remains occasional
n town parks (and some hedges/belts).

APPEARANCE Shape Tall-domed on often *straight, clean main limbs*. Most trees have dark, quite sparse foliage. **Bark** Grey; *regular* criss-cross ridges. **Leaves** Glossy above (but rough on low sprouts and the *frequent suckers*) and hairy *only in tufts under the vein-joints*; large (to 15 cm), on 15 mm stalks; the margin more often than in other elms (except 'Plantijn', p.134) curves in to meet the first vein *on the 'short' side*. **Fruit** With seed near wing's centre.

COMPARE European White Elm and 'Commelin' (p.134). Some wild hybrids (see p.130) are similar.

VARIANTS 'Dampieri Aurea' ('Wredei'), rare as a mature tree, is still being quite widely planted, as it shows good DED-resistance. Straggling, erect limbs make a broad column/funnel, to 16 m. Leaves small, rather round; jaggedly toothed and very crumpled, but soon smooth, *shiny* (cf. 'Louis van Houtte', p.131) and almost hairless: *brilliant rich gold* especially in later summer. Branches can revert to the otherwise almost extinct 'Dampieri' (similar, but with dull dark green leaves).

Dutch Elm

Ulmus × hollandica 'Hollandica'

('Major') A much planted Hybrid Elm (see p.130) since 1680; very occasional suckers remain.

APPEARANCE Shape A straggling, dark dome (once to 43 m); strong shoots corky-winged. **Bark** Brown; cracking into *smaller* shallow plates than English Elm's. **Leaves** Often buckled like English Elm's but much longer (to 15 cm); adult ones more or less smooth (cf. Coritanian Elm, p.133); downy only under the midrib. **Fruit** Seed *touches the fruit's notched margin.*

OTHER TREES *U.* 'Commelin' (Holland, 1940; Huntingdon × Smooth-leaved Elm) is a rare but often DED-resistant street tree. Steep branches with *dark grey*, closely ridged bark make a narrow, rather untidy and open crown, to 22 m (cf. 'Plantijn', opposite). Shoots matt brown; leaves slightly smaller and more oblong than Huntingdon Elm's; scarcely showing the curvature of the short side onto the first vein.

U. × hollandica 'Groeneveld' (Holland, 1963) is one of a group of newer, largely DED-resistant hybrid elms (with 'Dodoens', 'Lobel', 'Plantijn'). It is locally occasional: *columnar* on *sinuous* erect branches, then widening; bark smooth at first (cf. 'Dodoens'). Leaves about 8 cm, glossy above; *finely downy* beneath; fruits very freely.

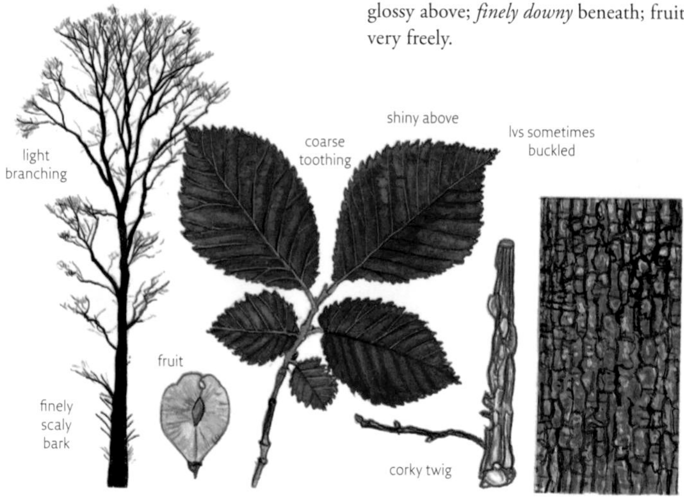

U. 'Dodoens' (Exeter Elm × Himalayan Elm, *U. wallichiana*; Holland 1973) is rare. Rather open, on steep branches; bark smooth and silvery at first (cf. 'Groeneveld'). Leaves about 10 cm; blackish, glossy and *deeply toothed*; hairless except for tufts under the vein-joints (cf. Dutch Elm).

U. 'Lobel' ('Dodoens' × U. × hollandica Bea Schwarz'; Holland 1973) is locally frequent. Narrowly oblong/funnel-shaped in youth (cf. 'Plantijn'); steep, straight main limbs; stiff shoots spread at narrow angles; bark thickly ridged then square-cracking. Leaves about 8 cm; glossy, blackish, almost smooth above, with a thick dark margin; only 2 mm of basal asymmetry (cf. Asiatic elms, p.136–7).

U. 'Plantijn' (Holland 1973; mostly Smooth-leaved Elm) is rare. Funnel-shaped laxer than 'Lobel'; cf. 'Commelin'). Leaves about 9 cm; slightly rough above; drifts of white hair under the vein-joints; slightly pie-crust margin has big teeth with up to four secondary teeth and curves in to meet the first vein on the 'short' side (like Huntington Elm, p.133).

European White Elm
Ulmus laevis

(Fluttering Elm; *U. effusa*) E France to the Caucasus. Very rare in the UK as an old wild tree (possibly native).

APPEARANCE Shape A billowing dome: often with fine sprouts and small burrs up the branches. **Bark** As Wych Elm's (p.128); may scale more finely. **Shoots** Downy at first. **Buds** *Orange*-purple, *long-pointed.* **Leaves** Long (to 13 cm; up to 19 vein-pairs), broadest below halfway; very asymmetrical at the base, the margin (with big, *very hooked* double teeth) sometimes curving in to meet the first vein *on the 'long' side*; rich green and quite shiny but slightly rough above, and often downy beneath; stalk 15 mm. Unlike the commonest similar elms', the *main veins very seldom branch.* **Flowers** *Long-stalked;* seeds *fluttering on their stalks; hair-fringed, with 2 converging horns* at the tip.

COMPARE Huntingdon and Belgian Elms (p.133). Only flowers and fruit (March–May) positively differentiate from Hybrid Elm variants.

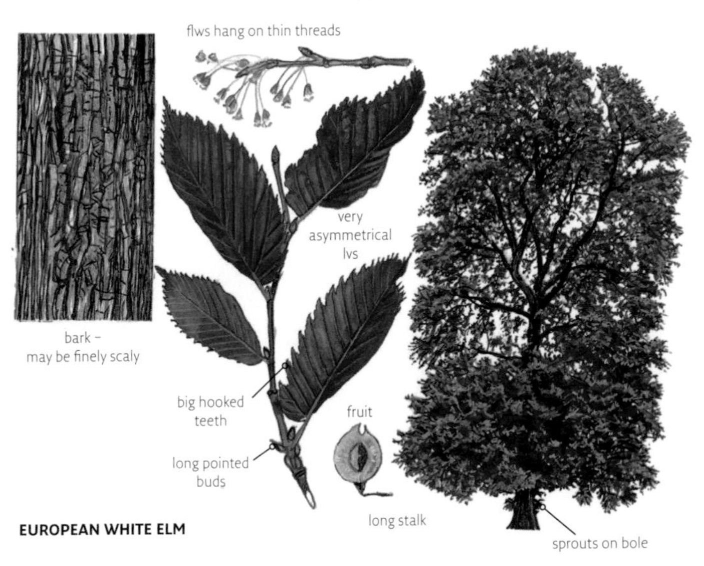

EUROPEAN WHITE ELM

Chinese Elm
Ulmus parvifolia

(Lace-bark Elm; *U. chinensis*) E Asia; S Japan. 1794. Rare. Often resists DED.
APPEARANCE Shape A beautiful tree, still green in late autumn. A dainty dome of narrow, blackish leaves; to 14 m. **Bark** Chocolate-brown, with coarse flaking ridges; or greyer and smoother between *orange scales*. **Shoots** Minutely grey-felted. **Leaves** 2–6 cm, rounded at the base with about 2 mm of asymmetry, and with blunt, *simple* teeth; upper side glossy or slightly rough; some hairs under veins. **Flowers** *In autumn*.

Siberian Elm
Ulmus pumila

(*U. microphylla*) N and E Asia. *c.*1860. Rare, but usually DED-resistant so now a little more planted.
APPEARANCE Shape *Broad, irregular and untidy*: curving branches from a short, usually slanted bole (cf. 'Sapporo Autumn Gold'); long, lax, often weeping shoots, like ostrich feathers (cf. Lock Elm, p.132, and 'Regal', above). To 20 m. *Fresh or pale green* healthy foliage in summer. **Bark** Willow-like: a *very coarse network* of scaly brown ridges. **Shoots** Soon hairless. **Buds** Small, glossy brown. **Leaves** Small (6 cm), slender and *hairless* (sometimes with tiny tufts under vein-joints), and more or less *symmetrical at the base*; up to three or four secondary teeth on each jagged main tooth; stalk 1 cm, finely downy.

Japanese Elm
Ulmus japonica

Japan (where always rare); in a few collections and beginning to be planted more widely for its DED resistance.
APPEARANCE Shape Usually seen in the UK – from the sapling stage – as a *very wi low*, deep-green dome of long, straggling, ostrich-plume shoots. **Bark** Grey-brown; very scaly ridges. **Shoots** Pale, typically downy (hairless in some planted trees); sometimes corky-winged. **Leaves** 3–10 cm typically scrubby above and downy beneat (but smooth, hairless and shiny in some planted UK trees), with asymmetrical base stalks 15 mm.

Ulmus 'Sapporo Autumn Gold'

(Japanese × Siberian Elms; Wisconsin, 1973) Frequent: the *generally planted* DEI resistant elm of the 1980s, though scarcely in commerce by 2000. The single clone is readily learnt by its jizz.
APPEARANCE Shape Asymmetrically and jaggedly broad-domed (to 19 m so far); *light branches rise from a very short, slightly slanted bole*; dainty and *fresh-green* from early spring, with many luxuriant, long but not drooping shoots. **Bark** Scaly criss-cross *brown* ridges; *orange* fissures. **Shoots** Finely downy. **Leaves** 4–9 cm (smaller an slenderer than the new hybrid elms' on p.134); *slightly rough* but glossy above, wi some down underneath; only about 2 mm of basal asymmetry (cf. 'Lobel', p.134).

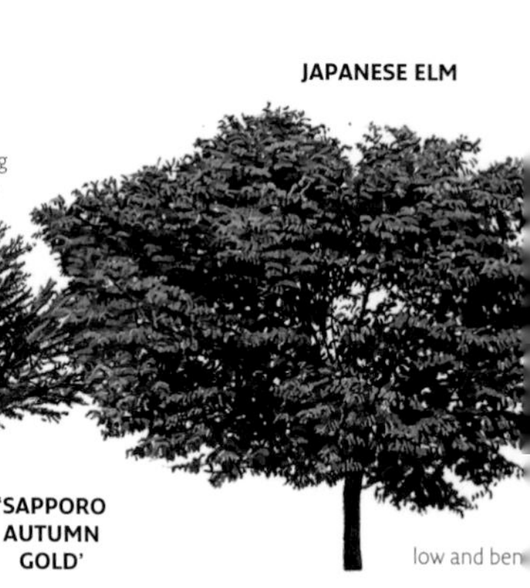

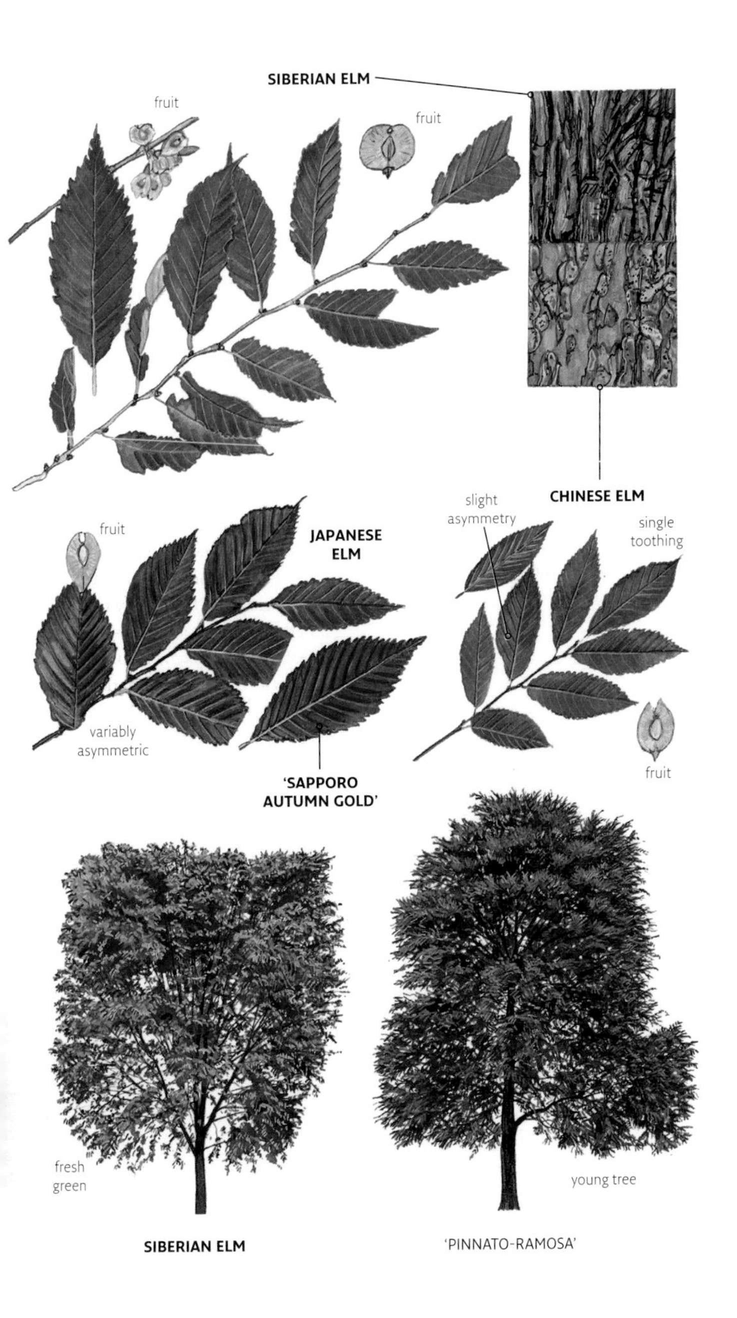
SIBERIAN ELM
fruit
fruit
CHINESE ELM
fruit
JAPANESE ELM
slight asymmetry
single toothing
variably asymmetric
'SAPPORO AUTUMN GOLD'
fruit
fresh green
young tree
SIBERIAN ELM
'PINNATO-RAMOSA'

Zelkovas (six species) are trees related to the elms. They can also succumb to Dutch Elm Disease (see p.128). (Family: Ulmaceae.)

Things to Look for: Zelkovas

- Leaves: What shape are they at the base? What shape are their teeth? How long is the leaf-stalk?

Caucasian Elm
Zelkova carpinifolia

(*Z. crenata*) Iran; Georgia; Armenia; E Turkey. 1760. Rather occasional; rarely naturalizing by its prolific suckers.
APPEARANCE Shape Usually a giant '*witch's broom' of steep stems from a 2 m bole with big, rounded flutings*, to a slender, slanted top (35 m), but sometimes 'tree-shaped' and occasionally bushy on sinuous, steep stems; rich, dark masses of little leaves. The 'witch's broom' shape (scarcely seen in the wild) makes big trees vulnerable to windthrow. **Bark** Buff-grey, remaining smooth but with a few *orange crumbling patches*. **Shoots** Slender, green/brown, hair[illegible] **Buds** Small, blunt: elm-like, but a brighter dark red. **Leaves** Hard, to 10 cm; 9–11 big rather *rounded* (but sharp) teeth each side; scattered scrubby elm-like hairs above and a softer down underneath, where stiff hairs also radiate from the main veins; stalks onl[y] 3–5 mm. **Fruit** (seldom seen) A green pea-sized nut.
COMPARE Chinese Zelkova; Macedonian Oak; Hornbeam 'Fastigiata' (p.107).
OTHER TREES Cut-leaved Zelkova, *Z.* 'Verschaeffeltii', is an obscure, daintily tree-shaped variant in a few big gardens, its

aves with *big outward-curving iangular lobes*. The bark, unlike :her zelkovas', may grow ıgged dark brown ridges. Cretan Zelkova, *. abelicea* (*Z. cretica*), endemic to Crete and ne of Europe's most ndangered trees. Bushy lants have grown in a few K collections since 1924: the leaves :rowded on downier shoots) are tiny .–4 cm), with only three to six pairs of eth.

:eaki
elkova serrata

Z. acuminata) Japan; Taiwan; Korea; NE :hina. 1862. Locally quite frequent as a uietly classy younger tree.
PPEARANCE Shape The commonly lanted clone quickly grows a *broad but raceful*, light-limbed dome, on a short, :raight, *smoothly rounded bole*, with anging fresh-green foliage; amber/pink in ıtumn. To 26 m; rarely a giant bush. **ark** Grey, with a few fine orange flakes; ıaggy plates may develop after 80 years. **Shoots** Hairless by autumn. **Buds** As Caucasian Elm's. **Leaves** *Slender, long-pointed*, to 12 cm, with 6–13 pairs of big *curved triangular teeth*; hairy only under the main veins; stalk 5–10 mm. Many trees grow odd sprays of miniaturized foliage.
VARIANTS 'Green Vase' is a recent erect selection.

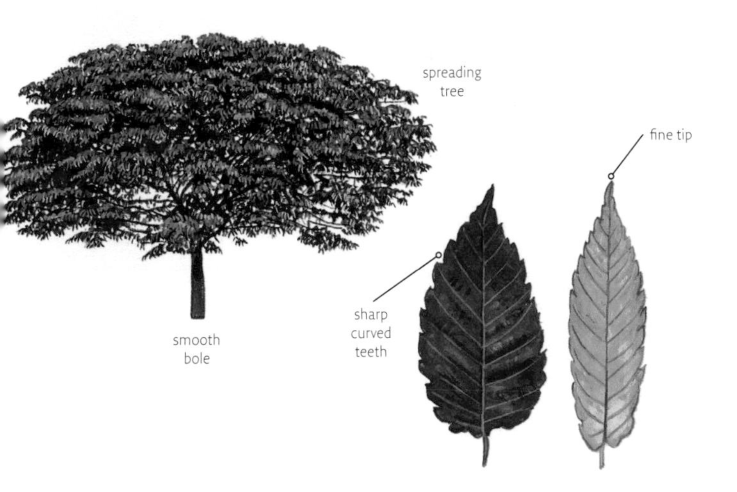

FIGS AND MULBERRIES

Moraceae is a predominantly tropical family (including 2000 figs). Members of the family have a latex-like sap.

Fig
Ficus carica

E Mediterranean and W Asia, but long grown in the UK. The hardiest member of a huge tropical genus. Frequent, and naturalizing from pips in rubbish, reaching tree size in the warmest areas; abundant along some rivers in N England (notably the Don at Sheffield) where industrial use of cooling water has historically lifted water temperatures. The sap (in sun) is irritant, especially to the eyes.

APPEARANCE Shape Often bushy or leaning; very gaunt in winter, with sturdily upcurving twigs; sometimes a suckering mass. In London a few sturdy trees have reached 13 m. **Bark** Elephant-grey and smooth. **Shoots** Thick and knobbly, green/grey. **Buds** Yellow-green, long-pointed, to 15 mm. **Leaves** Very leathery, with a sour, minty aroma; to 30 cm, variably lobed (rarely unlobed); shiny but rough and hairy above; downy underneath. **Flowers, Fruit** Dioecious, but male trees ('caprifigs') are probably not grown in the UK. Female flowers are produced on the inside of the young fig (which is a modified shoot-tip with a hole at the end, and is pollinated in the wild by tiny wasps). Most named clones are self-fertile. In Britain the common one is 'Brown Turkey', ripening its figs in their second year, given enough warmth.

Black Mulberry
Morus nigra

Probably of W Asiatic origin, but long cultivated across Europe. Quite frequent in warm areas in cottage gardens, old parks etc.; rarely naturalizing.

APPEARANCE Shape A low, dense, twiggy

ome on twisty branches from a crooked unk, to 12 m; late-flushing then singularly ıxuriant through summer. Mulberries were ften propagated by planting a 'truncheon' r large log – this sprouts but quickly rots om the sawn top, making young trees look ncient'. In fact there is no hard evidence ıat this is a long-lived tree; it certainly rows fast. **Bark** Orange-brown: scaly criss-oss ridges and some *bright, crumbling ıtches*; many big burrs. **Shoots** Stout, reyish, with some harsh hairs; strung with onspicuously *broad, sharp, purple buds* (cf. Vych Elm, p.128). **Leaves** 8–12 cm; thick; ark and *glossy but rough above* nd hairy, especially beneath; eart-shaped, but ariously dissected n sprouts and aplings). **Fruit** ipens in high ummer: avidly onsumed by nyone who oes not have worry about ıe incriminating lood-stains.

OMPARE alian Alder).105); Dove ree (p.228). White Mulberry (right): jizz ery different.

White Mulberry
Morus alba

China, but long cultivated in Europe: the favourite food-plant of the silkworm. Very occasional in warm areas of the UK.
APPEARANCE **Shape** An *open, upright dome of fine willowy shoots*; to 11 m. Often slender and straight-trunked. **Bark** Dull fawn-grey; a network of stringy but shallow ridges. **Shoots** Grey, soon smooth; small, sharp buds. **Leaves** *Pale fresh green*, flat and quite flimsy; nearly *smooth* and glossy above, but hairy under the veins. **Fruit** Pink/whitish: sweet but with little flavour.

MAGNOLIAS

Magnolias (about 80 species) have large 'primitive' flowers (landing pads for 'primitive' beetles of dubious aeronautical prowess). Their petals are not differentiated from sepals and are called 'tepals'. The leaves, always untoothed, are often enormous. Huge silky flower-buds adorn their shoot-tips; their berries dangle when ripe on silky strings from bright, knobbly 'cucumbers', before being blown free. Most are long-lived but of relatively recent introduction. (Family: Magnoliaceae.)

THINGS TO LOOK FOR: MAGNOLIAS

- Bark: How rugged is it?
- Shoots: What colour? Aroma when bruised?
- Buds: Are the leaf-buds hairy?
- Leaves: Are they evergreen? Pointed? Wrinkled? How long are they? How downy beneath? How far up are they broadest?
- Flowers: Are they erect or nodding? Scented? How many tepals and how wide? What colour?

MICHELIA

KEY SPECIES ✪

Southern Evergreen Magnolia (below): evergreen. **Cucumber Tree** (opposite): big leaves; flowers among them in early summer. **Campbell's Magnolia** (p.144): big leaves; flowers before them in spring. **Willow-leaved Magnolia** (p.145): small leaves; small white flowers before them in spring. **Saucer Magnolia** (p.146): bushy; tall, erect flowers before the leaves from an early age.

Michelia

Magnolia doltsopa

(*Michelia doltsopa*). E Himalayas to W China. *c.*1918. Larger gardens in mild areas. The hardiest of a big E Asian group of evergreen magnolias which flower from old wood, not at the shoot-tips.

APPEARANCE **Shape** Densely and rather bushily upright, to 20 m. **Bark** Grey; a few cracks with age. **Leaves** To 18 cm; *quite shiny above*, like a lustre-finish photograph; *silver-bloomed beneath* with fine down at first; rusty hairs persist under the veins. **Flowers** *Wreathing the shoots*, from rusty-silky buds, in early spring: creamy-white, to 10 cm; with a powerful, sweet, shaving-cream scent.

Southern Evergreen Magnolia

Magnolia grandiflora

(Bull Bay) Coastal SE USA. 1734. Quite frequent in warmer parts – often against ol[d] house walls.

PPEARANCE **Shape** Irregularly and stiffly omed; to 12 m. **Bark** Grey; big shallow cales slowly develop. **Shoots** Fawn-woolly. eaves Slender, to 25 cm; glossy above; range wool coating the paler under-leaf ubs thin through the year. **Flowers** Little y little from midsummer to late autumn. ichly scented; 9–15 huge tepals.
ARIANTS Many older trees in the UK are Exmouth': upright, its narrow leaves only *arsely* woolly and flowers with 18 tepals such distinct clones developed because uttings flowered after fewer years than lants raised from seed); 'Goliath' is bushy: *ig broad leaves, almost hairless beneath*; owers (from an early age) 30 cm wide.

OTHER TREES *Magnolia delavayi* (SW Yunnan, China; 1900; occasional in warm areas) has a fawn-grey bark with *close, corky ridges* and huge, broad leaves (to 35 cm), matt above and *silver-grey beneath*, with fine down. The flowers only open at night.

Cucumber Tree ✪
Magnolia acuminata

Ontario to Florida. 1736. Very occasional: older gardens in warmer parts.
APPEARANCE Shape Conic; then a leafy dome to 25 m. **Bark** *Orange-brown*, soon with shallowly scaly *ridges*. **Leaves** To 22 cm; pointed and broadest below halfway; bright green above; pale and finely downy beneath. **Flowers** 5–10 cm, green-yellow, rather lost among the leaves at the start of summer. **Fruit** Erect 7 cm 'cucumbers', often deformed, shocking pink then red.
VARIANTS Yellow Cucumber Tree, ssp. *cordata* (var. *subcordata*; SE USA; very rare to date), has flowers *bright yellow inside*. Usually bushy, with a more finely scaling bark and broader, darker, glossier leaves, long-hairy underneath.

Campbell's Magnolia ✪

Magnolia campbellii

(Pink Tulip Tree) Himalayas to W China. *c.*1870. Very occasional in milder areas: the most widely grown of the spectacular Asiatic spring-flowering tree species, but, as seedlings often take 20–30 years to begin to flower, hardly ever in smaller gardens.

APPEARANCE Shape Conic then openly domed, to 23 m, or on several straight trunks from the base; very wind-firm. The slender limbs run straight between sudden, sharp angles. **Bark** Typically grey with a few distant fissures; rarely rugged with close, buff, corky scales. **Leaves** Big (to 30 cm) and broad; rather matt; variably silky-hairy beneath. They are usually *oval with a small, sharp tip*, but can be long-tapered at the base like the other spring-flowering tree-magnolias'. **Flowers** Overwhelming the bare crown in *early* spring like a great flock of exotic birds – to 30 cm wide and typically a vivid *clear pink*. They remain more or less *erect* but at least the outer whorl of the *12–16 tepals* soon opens widely, flopping to the horizontal or beyond. The huge *ovoid* buds are vulnerable to frost-damage after their silky-hairy outer scales ha[ve] been shed, and each bloom only lasts a few days before the tepals bruise and fall.

VARIANTS f. *alba* has white flowers with a slight greenish cast: equally frequent in the wild but much less planted.

Ssp. *mollicomata* (Sikkim) is now more planted as a younger tree as it begins to flow[er] usually at 13 years from seed, and *a week later*, when the blooms are less vulnerable to late frosts: its flowers (*hairy-stalked*; fro[m] *oblong pointed buds*) are usually a pale, imperfect *purplish* pink but are shapelier

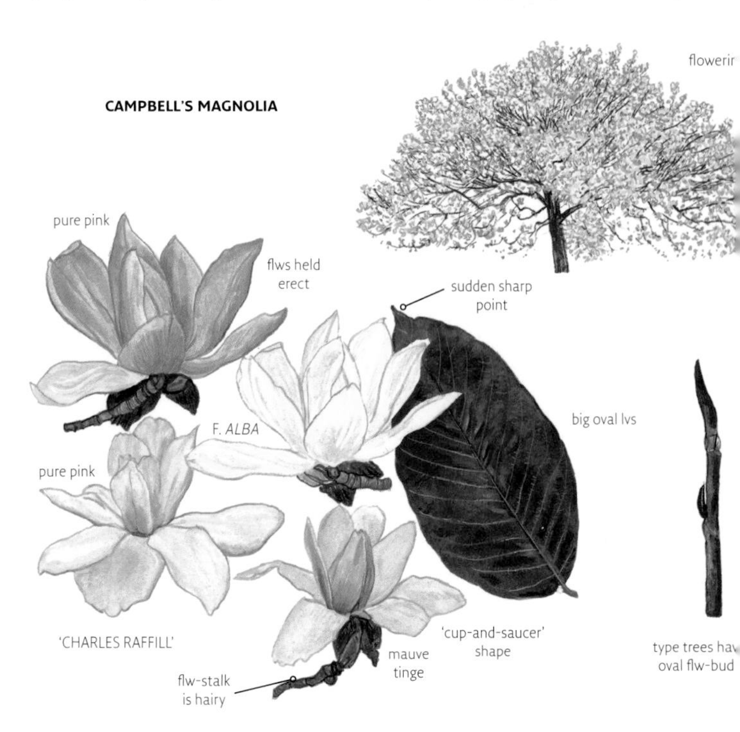

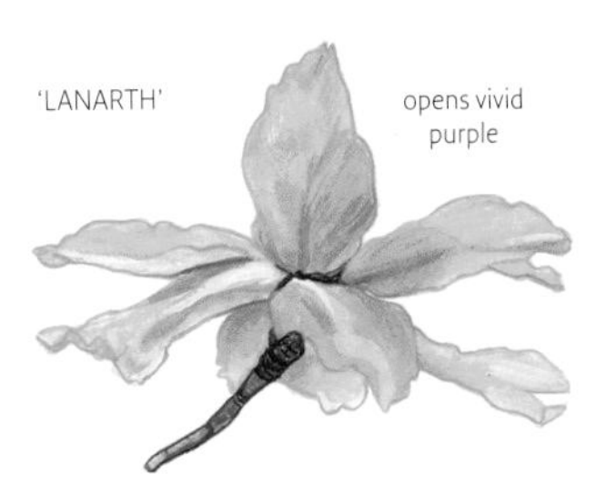

the outer ring of tepals ultimately spreads orizontally while the inner still curves nore regularly up, *like a cup on a saucer*).

'Lanarth' is a glorious wild-collected form f ssp. *mollicomata* with very *broad, thick eaves*. The 'cup-and-saucer' flowers open s early as the type's, an astonishing royal *nagenta-purple*.

'Charles Raffill' (Kew, 1946) is the least are of various garden hybrids between sp. *mollicomata* and the type: the flowers ombine the elegant 'cup-and-saucer' shape f the subspecies with the pure mid-pink olour of the best type trees, and are carried fter about 13 years from seed.

Willow-leaved Magnolia ✪

Magnolia salicifolia

apan. 1892. Occasional: larger gardens in varmer areas.

APPEARANCE **Shape** A slender-limbed onic/rounded tree, to 17 m; dense crisp domes of foliage, *red-flushing*. **Bark** Usually

buff; soon with small, ruggedly corky scales. **Shoots** *Slender, green*, hairless; *strong lemon-balm aroma* when bruised. **Buds** Flower-buds silky grey; *leaf-buds hairless*. **Leaves** The smallest of the tree-sized hardy magnolias': 8–14 cm and variably *slender, tapering* to a finely rounded tip; slightly *whitish* beneath, with a minute felt. **Flowers** Small (8 cm) snow-white nodding stars delicately overwhelm the tree a week after Campbell's Magnolia (opposite); six narrow, widely spreading main tepals; very sweetly scented.

COMPARE White Saucer Magnolias (pp.146–7): erect, *vase-shaped* flowers.

OTHER TREES Kobushi, *M. kobus* (N Japan and Cheju Do, 1865), is equally frequent. The stouter, *pale-brown* shoots have only a *slight* lemon fragrance when bruised; flower-buds *and* leaf-buds *very downy*; leaves darker, *green* beneath and *broader towards their tips*. There is usually a *leaf-bud at the base of the flower-stalk*. *M.* 'Wada's Memory' is one densely upright hybrid clone.

Star Magnolia
Magnolia stellata

(*M. kobus* var. *stellata*) Two sites in Japan. 1862. Abundant.

APPEARANCE Shape A twiggy bush (to 9 m). **Leaves** Slender, dark, undulant. **Flowers** Star-like, with *many* narrow white tepals; even on saplings.
OTHER TREES Loebner's Magnolia, *M. × loebneri*, is the hybrid with Kobushi (Germany; by 1910). Densely but delicately bushy, with stiff, straight twigs (*slight* lemon scent); now occasional (some small gardens) in various clones: solid in early spring, from an early age, with stars of many, spreading tepals. Leaves to 13 cm, from hairless buds.

'Merrill' (1939) is sturdily domed: its 15 cm flowers have up to 15 *broad*, pure white tepals.

'Leonard Messel', the most popular, is perhaps the loveliest magnolia: starry, spider 12 cm flowers, the 12 narrow tepals flushed pink (almost *lilac* at a distance). Slow, slender and soon flat-topped.

'Ballerina' (Illinois, by 1970) has *up to 30* tepals, white with a basal pink flush.

M. × proctoriana (the hybrid with Willow-leaved Magnolia, 1928) is a rare, dainty, small-domed tree (to 9 m): slightly hairy leaf-buds; dark leaves slenderer than Loebner's Magnolia's; starry flowers with six (or sometimes up to 12) tepals, white except for a faint basal pink flush.

Saucer Magnolia ✪
Magnolia × soulangiana

A hybrid of Yulan (above) with the more shrubby Japanese Lily-flowered Magnolia, *M. liliiflora*. The commonest magnolia, abundant in small gardens in a range of named and unnamed clones.
APPEARANCE Shape Untidily low-domed, to 13 m, or scraggy; usually a bush. **Bark** Smooth, grey. **Shoots** Grey-brown, with silky buds (leaf-buds 1 cm; flower-buds 2 cm). **Leaves** To 18 cm, broadest just

above halfway; often downy only under the midrib; flushing yellowish. **Flowers** Erect and vase-shaped, from an early age; nine white/pink/purple-stained tepals falling soon after they spread; spectacularly in early spring *then little by little through summer.*
VARIANTS Named clones include:

'Lennei': big flowers white inside, pale purple-pink outside, with a good autumn crop. Large, *broad* leaves, to 20 cm. 'Lennei Alba' is a seedling with pure white flowers.

'Brozzonii': large, *late* white flowers, purple-stained at the base; dark, matt leaves.

'Rustica Rubra': small flowers, mauve-red outside; dark, broad leaves; vigorous, rather *gaunt habit.*

'Picture': tall flowers, white within but with a central purple-pink stripe outside, and purplish at the base – found in a Japanese garden around 1930 by Koichiro Wada.

'San Jose' (1940, from California): creamy-white flowers, richly flushed pink.

'Verbanica': narrow tepals, usually a uniform mauve-pink outside; very late.

Tulip Tree
Liriodendron tulipifera

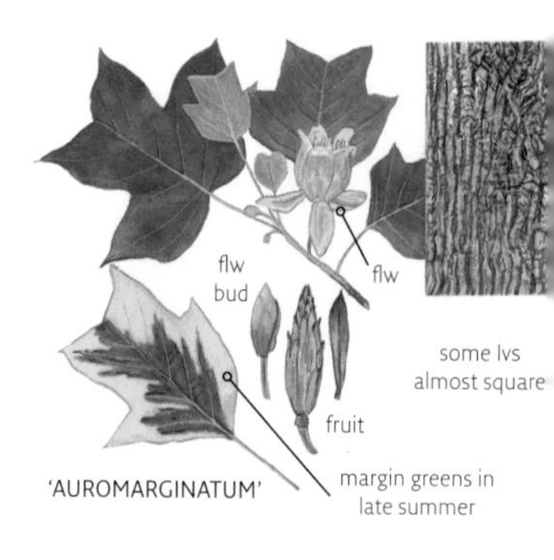

(Yellow Poplar) E North America: Nova Scotia to Florida. Quite frequent in warmer areas.
APPEARANCE Shape Towering, to 36 m; often rather twisted shoots and stems but sometimes starting off spire-shaped (like the best wild trees, which can have straight trunks 30 m long). Dense, rich-green summer foliage; reliably butter-yellow in autumn. Long-lived – one probable 1685 original thrives in Surrey – but easily broken by storms. **Bark** Pale grey: short, close, rather sharp criss-crossing ridges. Oldest trees more bronze and rugged; strange, ragged flanges. **Shoots** Greenish, bloomed lilac at first. **Buds** *Stalked and smoothly flattened, like a beaver's tail.* **Leaves** Deeply waisted on sprouts (like the adult leaves of Chinese Tulip Tree); occasionally with an extra pair of lobes; smooth and variably silvery beneath with a covering of minute waxy warts (papillae). (Juvenile plants have almost square leaves with no side lobes; the rare f. *integrifolium* retains this foliage.) **Flowers** Green-and-orange 5 cm tulips are abundant in June after a warm season but tend to be lost in the foliage. The 5 cm seed-heads last through the winter, biscuit-brown.
VARIANTS 'Fastigiatum' is spire-shaped then rather broadly columnar, on twisting, *vertical branches*, and remains rare.

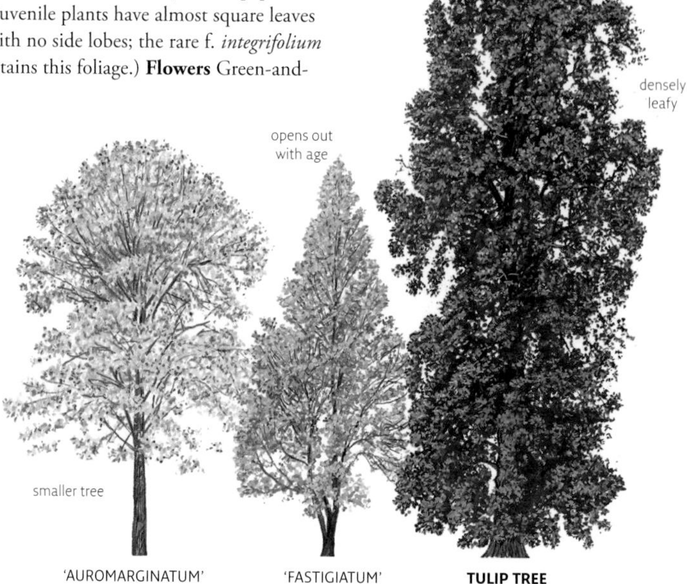

ureomarginatum' is rather rare but vigorous; e soft yellow leaf-margin fades to pale een through the summer and by autumn not obvious. 'Aureopictum' (much rarer) is yellow blotches at its leaves' centres.

hinese Tulip Tree
riodendron chinense

China to N Vietnam: discovered in 1875 another example of a genus common the two most diverse populations of mperate trees (E N America and China). 901. Still rare, less vigorous in the UK than ulip Tree (opposite); to 25 m.

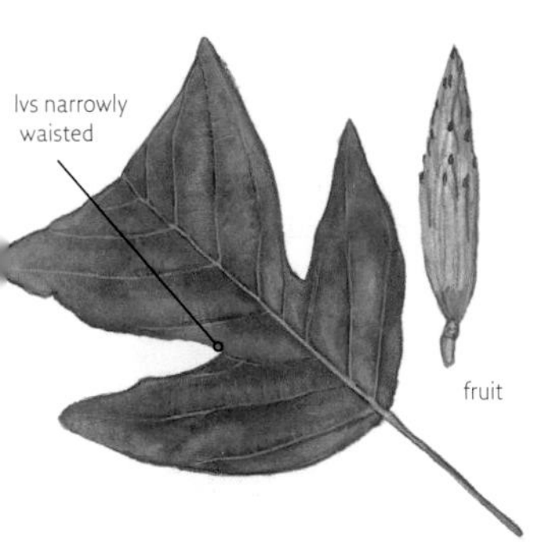

PPEARANCE Difficult to differentiate ith confidence from Tulip Tree. **Bark** Can ecome brown and rather finely craggy at a ounger age. **Shoots** More strongly white-loomed. **Leaves** Flushing briefly *purplish; lways* elegantly and strongly waisted (like ulip Trees' sprout-leaves); a denser covering f papillae makes them somewhat whiter nderneath. **Flowers** Generally greener and few weeks later than Tulip Tree's.

Katsura
ercidiphyllum japonicum

China; S Japan (an endangered giant). 1865. ather occasional; happiest in mild, humid reas. (Family: Cercidiphyllaceae.)
PPEARANCE Shape Daintily rounded-conic, to 25 m, often on several straight stems (Chinese trees, var. *sinense*, are more often single-trunked). Prominent paired buds; sweeping minor branches closely strung with neat little leaves, even inside the crown. **Bark** Pale brown/grey; long shaggy strips. **Leaves** In *opposite pairs* (alternate at the shoot-bases), finely round-toothed (the smallest scarcely so); to 10 cm; soon hairless. They flush pink and die off lemon and orange, with a *caramel scent* from high summer onwards which can lead you to distant trees. **Flowers** Dioecious: red tufts. **COMPARE** Judas Tree (p.196): similar leaf-shape.

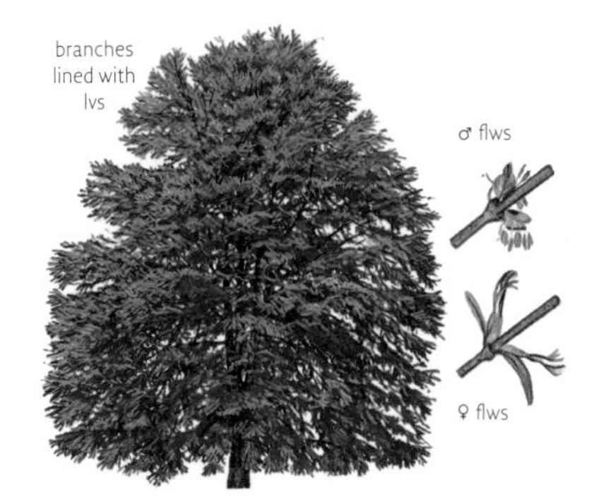

KATSURA

Sassafras ☠
Sassafras albidum

(*Sassafras officinale*) Ontario to Florida, and W to Texas. In Spain by 1560. A rare but stylish tree, confined to acid soils in warm areas. (Family: Lauraceae.)

APPEARANCE **Shape** A rather narrow dome of twisting branches: to 18 m but sometimes shrubby and gaunt; suckering freely. **Bark** Grey; close sharp ridges. **Shoots** *Slender; leaf-green* for some years; *buds green*. **Leaves** Glossy green above, silvery under (downy in var. *molle*); *strongly vanilla-scented* (but carcinogenic?). Unique *range of oven-mitt shapes*, though in old trees unlobed leaves are the norm. In autumn, yellow, pinkish red and rich orange. **Flowers** Usually dioecious. **Fruit** Female trees carry black 1 cm berries on long red stalks.

COMPARE In winter, Sweet Gum (opposite); green shoots distinguish Sassafras.

Bay 💧
Laurus nobilis

(Sweet Bay; Poet's Laurel) Mediterranean area. Long grown further N; abundant in milder areas; rarely naturalizing. 'Bachelor' (*baccalaurus*) derives from 'laurel-berry': th graduate was wreathed with Bay.

wavy margin

bushy in colder areas

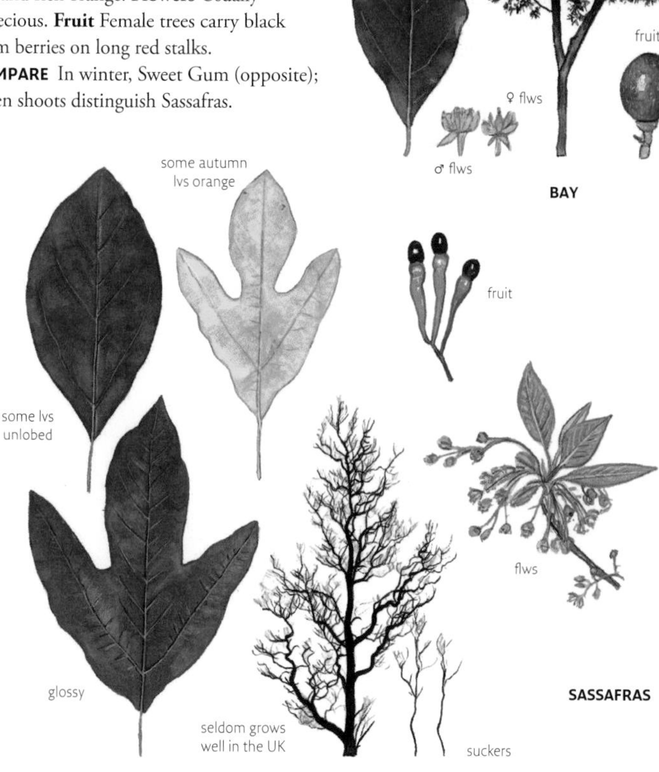

PPEARANCE **Shape** A dense evergreen pire to 20 m; bushier in colder parts. **Bark** mooth, black-grey. **Leaves** 5–12 cm, finely ong-pointed; *thin* but hard with crinkled dges and the odd tiny tooth; *hot, sharp fruity mell.* **Flowers** Dioecious; egg-yolk yellow in pring, from yellow globular buds prominent l winter. **Fruit** 12 mm black berries.
OMPARE Holm Oak (p.123): leaves woolly beneath. Strawberry Tree (p.234): regularly serrated leaves. Other evergreens (e.g. hollies) have thicker/glossier/less finely pointed leaves.
VARIANTS Narrow-leaved Bay, 'Angustifolia' (undulant leaves less than 2 cm wide), is a remarkably distinctive but very rare foliage plant.

Golden Bay, 'Aurea', has gold leaves through winter/spring; rare, though hardier than the type.

Hamamelidaceae is a diverse family of often winter-flowering trees and shrubs.

Sweet Gum
iquidambar styraciflua

American Sycamore) SE USA to Mexico – ften in swamps; planted commercially for he balsam in its sap (the medicinal liquid torax). 1681. A frequent garden tree in warmer areas.

PPEARANCE **Shape** Stiffly conic, then omed or irregular, to 30 m, with twisting, pcurving, often broken branches and (in winter) a *spiky* outline. **Bark** Grey-brown; nobbly, scaly ridges from the first. **Shoots** Young twigs (especially from suckers) can have corky wings (cf. English Elm, p.130; Field Maple, p.205). Saplings' short, spiky side-shoots with long-pointed buds suggest Aspen's (p.79) but are more forward-angled; buds are a clearer red/green. **Leaves** Five-lobed, finely toothed: maple-like but *alternate*; with small *tufts of hairs* under the vein-joints and scurf under the veins. Young trees can have three-lobed leaves (cf. Chinese Sweet Gum, below). Fresh, *pale green* in summer; splendid autumn colours – lemon-yellows, crimsons, saturated purples. **Fruit** Hangs through winter – 3 cm spiky balls, rarely seen in the UK.

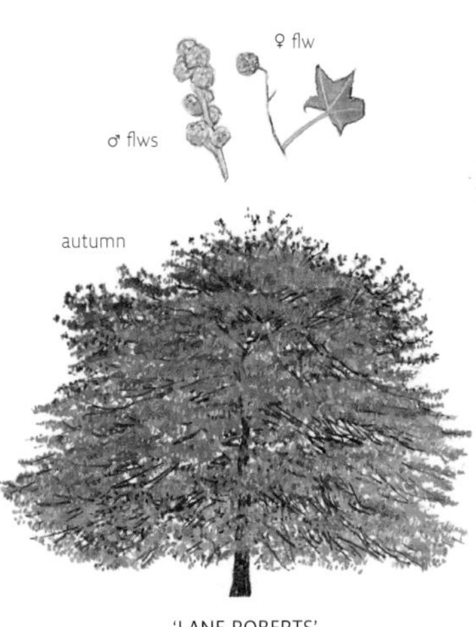

'LANE ROBERTS'

VARIANTS 'Variegata' ('Aurea') is very rare: leaves yellow-splashed; 'Silver King' (young plants; rare; sometimes also sold as 'Variegata') has clear creamy-white leaf-margins.

'Lane Roberts' and 'Worplesdon' are autumn-colouring selections, seen occasionally as younger, grafted trees. 'Worplesdon' (crimson/yellow) tends to be bushy, with small, narrowly lobed leaves, but 'Lane Roberts' (blackish crimson) grows a good straight trunk of relatively smooth bark.
OTHER TREES Chinese Sweet Gum, *L. formosana* (China, Taiwan; 1884), is rare; to 23 m. Bark grey, usually *smooth*; leaves very *matt*, more or less *3–lobed*; usually with some hairs across the underside. Autumn colours late, and less reliable.
Oriental Sweet Gum, *L. orientalis*, a large tree from Turkey and Rhodes (1750), remains bushy in the UK; very rare. Bright brown square-cracked bark; smaller matt leaves *completely hairless* beneath.

Persian Ironwood
Parrotia persica

Forests S of the Caspian Sea. 1841. Rather occasional: town parks, larger gardens.
APPEARANCE Shape *Long, gaunt branches typically arch to the horizontal* from a dense central crown on a (very) short, muscularly fluted trunk. To 15 m. **Bark** Fine plates flake in *cream, grey and orange*: cf. Strawberry Dogwood (p.234). **Shoots** Wit short, stiff hairs. **Buds** Purplish with short, stiff hairs. **Leaves** Glossy, to 12 cm; odd teeth/waves (like the shrubby witch-hazels'). Yellow and red autumn colours start early a the branch-tips. **Flowers** Maroon stamen-clusters in late winter (lacking witch-hazels' yellow petals).

he 6–7 species of plane originate in North America and Asia. (Family: Platanaceae.)

ondon Plane
latanus × hispanica

P. × acerifolia) Probably a fertile hybrid f Oriental Plane and the American uttonwood, *P. occidentalis* (with clumsy, hallowly lobed leaves and scarcely grows n the UK; also called 'Sycamore' in North merica, while in Scotland this book's Sycamore' is sometimes called 'Plane'), riginating in Spain or S France (*c.*1650) ut long planted abundantly in S England; ery rarely naturalizing.

PPEARANCE **Shape** Long, straightish mbs rise from a buttressed trunk generally *lean for some metres*; smaller branches very wisting. To 44 m; oldest trees thriving

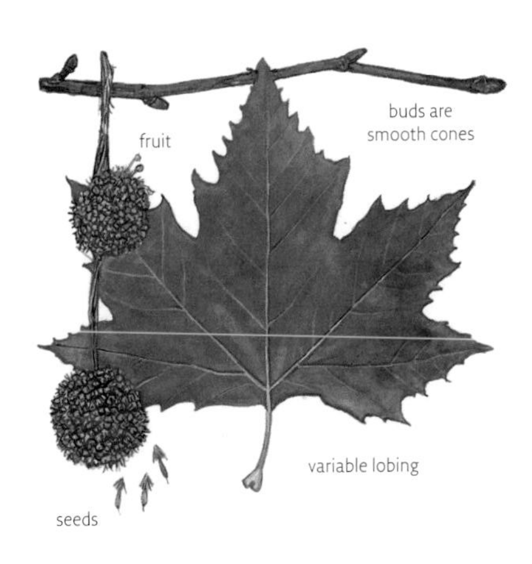

at 320+ years. **Bark** Shallowly *scaling in greys and creams* – but see variants. **Shoots** Green/brown, scurfy at first, the end bud missing. **Buds** *One-scaled red/green horns*, hidden in the leaf-stalks in summer and in winter rimmed by their scars. **Leaves** Thick; maple-like but *alternate*; typically three-plus teeth on the shoulder of each main lobe. *Hard, deep green*; flushing late and buff-woolly. **Fruit** 3 cm balls of hairy seeds (usually two to six per string), breaking up in spring – can bring on asthma attacks.

VARIANTS Several clones can be distinguished, though few are named. One (mostly eighteenth-century plantings) has a tall, duller brown trunk and leaves with long, often untoothed triangular lobes; one has big leaves and greyish basal bark cracking in rectangles. Another (which could be called 'Baobab Plane', and probably carries a viral infection), has disproportionately small, very twisting branches on a short trunk *extraordinarily*

swollen with rough burrs; its deeply cut leaves resemble Oriental Plane's but are darker, the lobes having *more* long teeth.

'Pyramidalis' (locally frequent) was christened before its crown had matured: it has straighter branches and is narrowly columnar at first, then broad and gaunt. The glossy leaves resemble Buttonwood's in being *shallowly three-lobed* (smallest ones almost round) and the bark is soon *dull brown and furrowed, with many burrs*. The 45 mm fruit-balls, *one to two per string*, are like Buttonwood's.

'Augustine Henry' is splendid but occasional: long bole and straighter branches clad in very clean, grey and creamy-white bark; sea-green leaves with drooping edges and *up to five regular long teeth* on the shoulder of each lobe; fruit-balls *one to three per string*.

'Suttneri' is brightly white-variegated but very rare.

Oriental Plane
Platanus orientalis

SE Europe, with ancient planted trees E to Kashmir. Long grown in the UK; now rather occasional, in warmer areas.
APPEARANCE Shape Generally broader than London Plane: to 30 m: short, often massively burred bole and limbs which can bend to the ground and layer. Typically *fresh green*, delightfully *feathery* foliage; pale bronzy-purple in autumn. **Bark** As London Plane; on old trees pale brown and rugged (but almost white in hot climates). **Leaves** With deeper finger-like lobes than London Plane's (especially narrow in 'Digitata'); normally with only one or two teeth on the shoulder of each main lobe; often ('Cuneata') narrowly tapered to the base. *Sweet, balsam smell* from the foliage is stronger than in London Plane. **Fruit** Like a typical London Plane's but smaller: three to six balls per string.

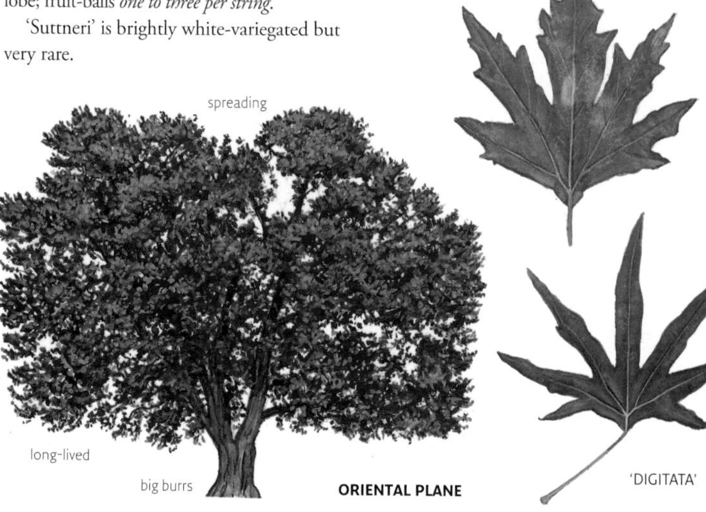

ROSE FAMILY TREES

Rosaceae is a family of 3000 herbs, trees and shrubs, many with edible fruits.

Himalayan Tree Cotoneaster
Cotoneaster frigidus

Himalayas. 1824. Quite frequent.
APPEARANCE Shape Often multiple, leaning stems with *many vertical sprouts*; sometimes a sturdy tree, to 15 m. **Bark** Pale fawn-grey, harshly *scaly*. **Leaves** *Flat*, untoothed, 6–12 cm, falling through early winter; whitish beneath with a close felt reduced to sparse down by autumn; *blunt*, a tiny downward bristle at some tips. **Flowers** White: 5 cm heads in early summer. **Fruit** *Deep red* 7 mm berries can last till spring.
COMPARE Medlar (p.160). Hybrid Cockspur Thorn (p.159): serrated leaves.
OTHER TREES *C.* × *watereri* covers a swarm of (occasionally naturalizing) hybrids with the evergreen *C. henryanus* and *C. salicifolius* (which have smooth grey barks); *veins impressed and leaf-margin downcurved*. *C.* 'Cornubia' (frequent; sometimes placed under *C. frigidus*) has the largest leaves (to 13 cm), almost hairless by late autumn,

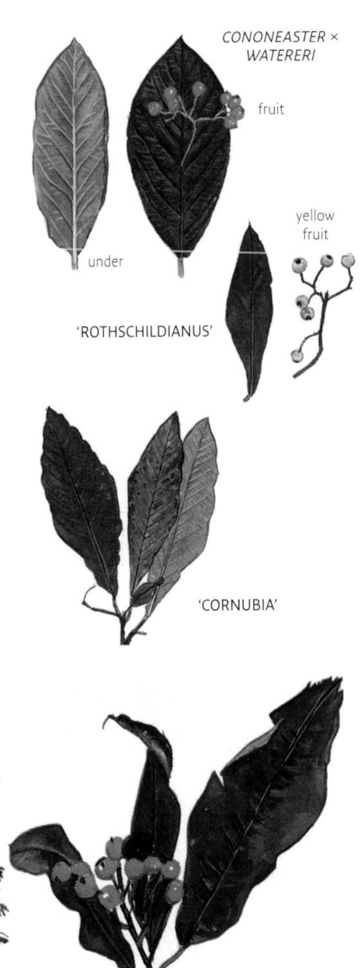

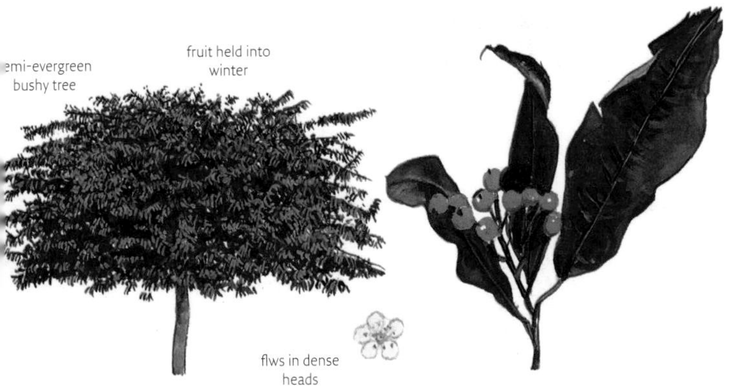

when they begin to drop; berries slightly larger (1 cm) and more vividly blood-red than Himalayan Tree Cotoneaster's. *C.* 'Exburiensis' has lemon-yellow and *C.* 'Rothschildianus' amber fruits; their smaller, more or less evergreen leaves remain woolly beneath.

Loquat

Eriobotrya japonica

(Japanese Medlar) China. 1787. Occasional in milder parts: in small gardens as often as big ones; rarely seeding.

APPEARANCE Shape Rounded, often flat and bushy; surfaced with disproportionately *large, dark, drooping evergreen leaves* (to 30 × 12 cm; smaller with age). **Bark** Blackish; distantly cracked with age. **Leaves** Glossy; with *impressed veins* and beige felt beneath; *coarsely toothed*. **Flowers** Hawthorn-scented, creamy: on 15 cm hea in autumn. The sweet, yellow, pear-shape 4 cm loquats ripen next summer (but seldom in the UK).

COMPARE Giant Photinia (below); Southe Evergreen Magnolia (p.142).

Giant Photinia

Photinia serratifolia

(*P. serrulata*) China and Taiwan. 1804. Ve occasional in bigger gardens in milder par

APPEARANCE Shape A densely surfaced evergreen dome, to 15 m. **Bark** Grey; smooth; some reddish scales with age. **Leaves** Large (to 20 cm); leathery; very *finely serrated*; white down under the midrib at first, then hairless. They unfold glistening brownish red (this is one parent of the ubiquitous, much smaller-leaved bush, P. × *fraseri*). **Flowers** In 15 cm, whit hawthorn-scented heads among the bright spring leaves; 6 mm red berries follow.

COMPARE Cherry Laurel (p.195): distant teeth. Southern Evergreen Magnolia (p.14

HAWTHORNS

awthorns (over 200 species) are small, usually thorny trees.

HINGS TO LOOK FOR: HAWTHORNS

Leaves: What shape? Are they downy above/beneath?
Flowers: How many anthers? and what colour? How many styles?
Fruit: How big? What colour? How many pips?

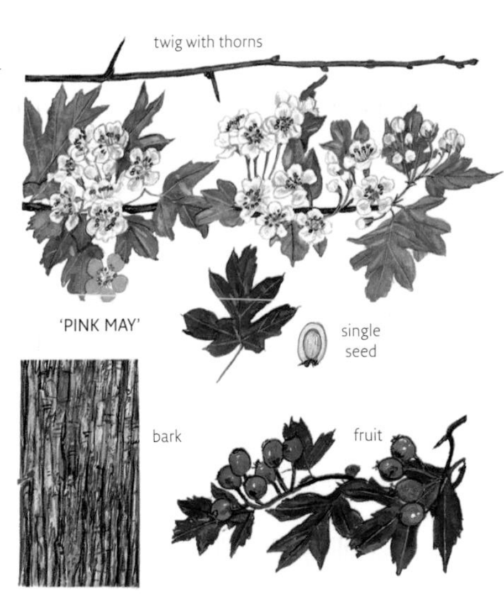

EY SPECIES 

ommon Hawthorn (below): deeply bed leaves. **Broad-leaved Cockspur horn** (p.159): unlobed leaves.

ommon Hawthorn 

rataegus monogyna

Quickthorn; Whitethorn; May) urope (including Britain and eland) to Afghanistan. Dominant in ost hedgerows; abundant everywhere xcept on poor sands.

APPEARANCE **Shape** A twiggy mass of stiff branches, to 15 m. Can be long-lived. **Bark** Brown; shallow, often spiralled, scaling ridges. **Shoots** Glossy green- or red-brown; many *1–2 cm rose-like thorns*. (The 'spines' of other 'thorny' trees only *tip bud-bearing side-shoots*; Hawthorns have these too.) **Leaves** (To 6 cm) Cut *at least halfway to the midrib* by lobes *toothed at their tips*; main veins *curve outwards*, with *tufts under the joints*. **Flowers** White (rarely fading red); *one style*. **Fruit** 'Haws' with *one pip*.
COMPARE Midland Hawthorn (below).
VARIANTS 'Fastigiata' is occasional: as nearly vandal-proof as any street tree, but no beauty.

'Pink May' has delicate pink flowers but is much rarer than the red cultivars of Midland Hawthorn. ('Paul's Scarlet' is colloquially called 'Pink May'.)

Glastonbury Thorn, 'Biflora' (very rare), is said to have sprung from the staff of Joseph of Arimathaea, planted on Christmas Day at Glastonbury in front of a sceptical heathen congregation; its *premature secondary crop* of leaves and flowers is seldom as early as December.

OTHER TREES Oriental Thorn, *C. orientalis* (*C. laciniata*), is the most planted of several SE European thorns whose similarly-shaped (but rather larger) leaves are softly *grey-downy*. Haws *18mm*, bright brick-red: very showy.

Midland Hawthorn

Crataegus laevigata

(*C. oxyacantha*; *C. oxyacanthoides*) W Europe, including S/central England: ancient woods/old hedges *on clay*.

APPEARANCE Shape Bushy in the wild, but odd park trees are as large as Common Hawthorns. **Leaves** Dark, slightly glossy, and lobed *less than halfway to midrib*; almost hairless; main veins straight or *curved upwards*. **Flowers** *two to three styles*. **Fruit** Haws with *two to three pips*: hybrids with Common Hawthorn (*C.* × *media*: as widespread as the species) have one to three. Some wild trees (f. *rosea*) open pink.

COMPARE Common Hawthorn (above): differences emphasized. Grignon's Thorn (opposite).

VARIANTS Many probably derive from *C.* × *media* (leaves lobed at least halfway to the midrib). 'Punicea Flore Plena' (frequent) has mauve-pink double flowers; some may be reversions from 'Paul's Scarlet' – an abundant sport (1858) of this old form, whose *double crimson* flowers fade whitish inside. 'Masekii' (rare) has double soft pink flowers. 'Plena Alba' has double white flowers (fading pink; occasional). 'Punicea' (very occasional) has single flowers of crimson petals and a white heart; 'Crimson Cloud' is similar.

ɪoad-leaved Cockspur ɪorn ✪

ɪtaegus persimilis 'Prunifolia'

prunifolia) A selection (by 1797) of ɪther North American thorn: the *glossy ɪl leaves* turn brilliant orange-red in ɪumn. Frequent in parks, streets and ɪdens.

PEARANCE Shape Broad and twiggily ɪned; to 9 m. **Bark** With very scaly, often ɪalling ridges. **Shoots** Hairless; thorns ɪm. **Leaves** To 8 cm, *never lobed; shiny*, ɪ finely hairy under the midrib. **Fruit** ɪrk red haws, 15 mm, adding (briefly) to autumn spectacle.

MPARE Snowy Mespil (p.179): leaves less ɪssy. Other thorns with scarcely lobed (but ɪrower) leaves include Hybrid Cockspur ɪorn.

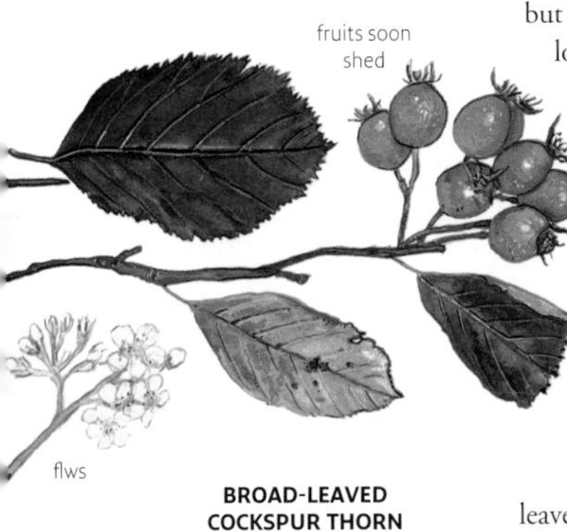

Hybrid Cockspur Thorn

Crataegus × lavallei 'Carrièrei'

(*C × carrierei*) A hybrid probably of Cockspur Thorn with the semi-evergreen Mexican *C. stipulacea*; by 1880. Frequent in parks and streets.

APPEARANCE Shape Picturesque: rather level, *dense* branch-systems; to 12 m. **Bark** Grey; very scaly. **Shoots** With long hairs but few thorns. **Leaves** *Long* (to 10 cm, the lowest third untoothed), glossy, *blackish*; some stiff hairs above and down underneath: they flush late, then look almost evergreen until they drop, still green, early in winter. **Flowers** Showy, in early summer, on woolly stalks. **Fruit** Dull scarlet, 18 mm; lasting until spring.

OTHER TREES Grignon's Thorn, *C. × grignonensis,* is another, rather rarer hybrid of *C. stipulacea* (*c.*1873), more upright and untidy. Shoots hairless; leaves much shorter (to 6 cm), paler and

less glossy, with *variable large teeth/small forward-pointing lobes* – on strong shoots they can resemble Midland Hawthorn's (p.158). Flowers in late spring; haws (some lasting all year) *bright* red.

Medlar
Mespilus germanica

SE Europe to Iran; long grown in the UK and (rarely) well naturalized in SE England; occasional in old gardens. Grafted often on hawthorn and sometimes on pear.

APPEARANCE Shape Low and *tangled*; spiny. (Old fruiting cultivars – 'Nottingha 'Royal', 'Dutch' – are thornless.) **Bark** Gr brown; long, scaly oblong plates. **Shoots** Densely white-hairy at first; prominent pa grey lenticels. **Leaves** To 15 cm, on *5 mm stalks, untoothed*; dull and wrinkly; dense hairs beneath. **Flowers** White, to 6 cm. **Fr** In Britain the 5 cm medlars are edible – h pear, half date – only when they have gon 'sleepy' in late autumn ('bletted').
COMPARE Himalayan Tree Cotoneaster (p.155).

Quince
Cydonia oblonga

(*Pyrus cydonia*) Long cultivated and rarel naturalized in the UK (probably W Asian Now very occasional (but a rootstock for many pears).
APPEARANCE Shape Low; often bushy. **Bark** Grey, *smooth;* then with some big, thin, rufous scales. **Leaves** Broad, to 10 c *untoothed*; woolly beneath when young. **Flowers** White/pink, 5 cm. **Fruit** The quinces (Portuguese *marmelo*, whence marmalade) are overwhelmingly fragrant, but inedible raw.

VHITEBEAMS, ROWANS & SERVICES

he *Sorbus* genus comprises 100 trees and shrubs including whitebeams (simple aves) and rowans (compound leaves).

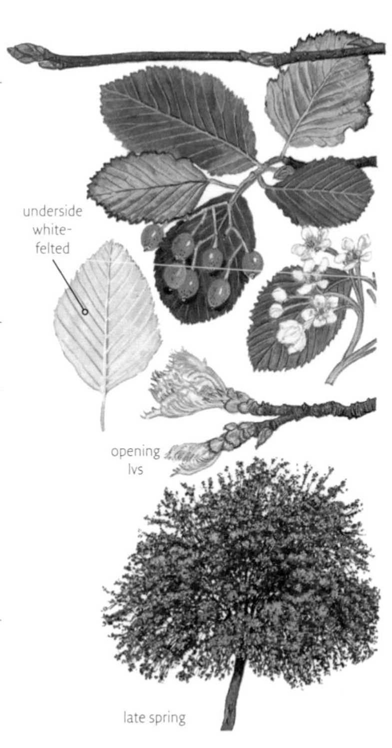

HINGS TO LOOK FOR: *SORBUS*

Bark: What is it like?
Buds: Are they hairy? Sticky?
Leaves: What shape? Are they hairy beneath?
Fruit: What colour? How big? How many seeds?

EY SPECIES 

ommon Whitebeam (below): oval leaves, hite-downy beneath. **Swedish Whitebeam** .164): lobed leaves, grey-downy beneath. **'ild Service** (p.163): lobed leaves, soon most hairless. **Bastard Service** (p.165): me leaflets at the leaf-base. **Common owan** (p.167): fully compound leaves.

ommon Whitebeam

rbus aria

and central Europe, including S England d Co. Galway; Morocco. Locally frequent the wild on chalk scarps and light sands; undantly grown everywhere.
PEARANCE Shape A narrow dome on ff, steep branches; to 23 m. Wild trees are ten multi-stemmed. **Bark** Grey; smooth, en with odd fissures. **Shoots** With white ool quickly shed; *brick-red* in sun, grey-een in shade. **Buds** Conic, to *15 mm; own and green* down-tipped scales. **Leaves** 12 cm, irregularly toothed/slightly und-lobed; 8–13 pairs of *close-set* parallel ins; stalks *slender*. They unfold erect and very, like magnolia blooms; the hairs ove are soon shed but the under-leaf ys startlingly white-woolly. **Flowers** In 7 cm-wide heads. **Fruit** Dull red fruit with two seeds, which birds soon pillage.
COMPARE Other whitebeams; *S. mougeotii* (p.165); *S. croceocarpa* (p.164); Pillar Apple (p.172).
VARIANTS 'Lutescens' (1892) is the most planted clone; erect then neatly narrow-domed to 14 m. The leaves (broad but not long) retain some *pale mealy down above* until late summer, giving the crown a steel-grey cast; berries grey-woolly until half-ripe.

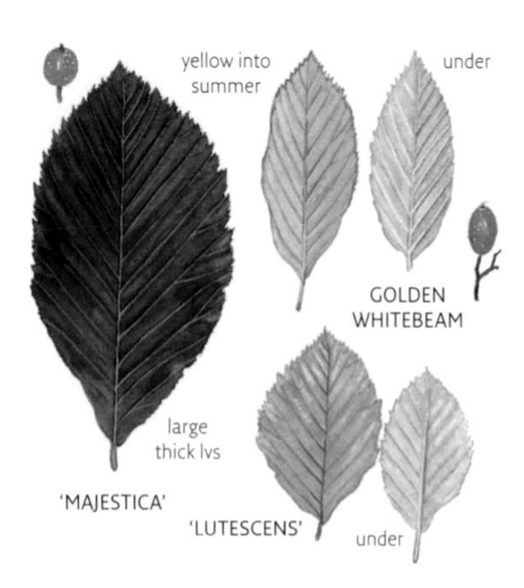

'Majestica' ('Decaisneana') is a group of frequent clones, to 20 m: long/broad leaves (to 15 cm), *dark* and glossy above, with up to 15 close vein-pairs (cf. 'Wilfrid Fox').

Golden Whitebeam, 'Chrysophylla', is rare: slender leaves pale yellow above, fading to olive.

OTHER TREES Cliff Whitebeam, *S. rupicola*, is the least rare of several British and Irish microspecies: limestone cliffs in Scotland, Wales, Ireland, the Pennines and Devon (also in Scandinavia and Estonia). Leaves *small (3–7 cm)*, the lower third narrowly tapered and untoothed, with very small, sharp lobes only round the *almost fan-shaped* top; fruits broader than long. Subtly distinct are *S. eminens* (Wye Valley and Avon Gorge), *S. hibernica* (central Ireland), *S. lancastriensis* (S Cumbria), *S. porrigentiformis* (N Devon, the Mendips, S Wales), *S. vexans* (N Devon) and *S. wilmottiana* (Avon Gorge). All are bushy in their typical habitats, and confined as planted trees to a few botanic gardens. *S. graeca* (Sicily E to Iraq; N Africa) covers a complex of similarly small-leaved trees; *S. umbellata* (Balkans to Crimea) covers another group with particularly squat, fan-shaped leaves, intensely white beneath and with only five to eight vein-pairs.

Himalayan Whitebeam
Sorbus vestita

(*S. cuspidata*) Himalayas. 1820. In many b gardens (some may be hybrids).

APPEARANCE Shape To 20 m; intensely steely-grey. **Bark** Slightly rougher than Common Whitebeam's: papery, grey-purp ridges. **Shoots** Thick, woolly-white. **Buds** Pink/green, blunt. **Leaves** Long (to 20 cm) and variably broad, stout-stalked and thick, yet with *only 6–12 distant vein-pairs* (cf. *S. croceocarpa*, p.164), the white felt beneath *plastered flat* as if with a tiny iron. **Flowers** Hawthorn-scented. **Fruit** *20 mm*, russet, with four to five seeds and ripening in *early winter*.

OTHER TREES *S.* 'Wilfrid Fox' is probably a hybrid (*c.*1920) with Common Whitebean named after the creator of the Winkworth Arboretum in Surrey and very occasional. Stiffly slender at first; to 14 m. Shoots

out, soon dark brown;
ıves oblong, inheriting
imalayan Whitebeam's
ɔut stalks/midribs and
illiant felt, with 12–15
ırs of *slightly incurving*
ain veins. (The *closer* veins
the Common Whitebeam
Iajestica', opposite, tend
curve outwards.) Fruit *to 20 mm, dull*
ange-brown.

S. hedlundii, in some collections, has *12–17*
ırs of veins with (on adult trees) *orange* hairs
neath them (in contrast to the white felt).

S. pallescens (China 1908; in some
llections, to 20 m) has pointed, narrowly
long leaves, to 12 × 5 cm (cf. Folgner's
ıitebeam), grey-green underneath by
tumn *as the white felt wears thin*; veins in
–13 pairs; fruits with two to five seeds.
ıe bark may be finely scaly.

WILD SERVICE

/ild Service ✪
rbus torminalis

:hequers Tree) S and central Europe,
:luding England N to Cumbria; Algeria;
: Caucasus; Syria. Locally frequent on heavy
ls, reproducing in Britain from suckers
uch more often than from seed (summers
: often too cool for seeds to ripen; plants
sed from seed can be *S.* × *vagensis*, p.164);
:ently more widely planted.
PEARANCE **Shape** A large tree (to 28 m);
aight twigs on heavy, twisting limbs. **Bark**
ey, then black-brown with close, scaly
long plates. **Shoots** Shiny grey-brown,
ɔn hairless. **Buds** *Green, round, like peas.*
aves Resemble only Italian Crab's but
larger and more sharply lobed; shiny
ıeath, with hairs soon confined to the
ns. Sometimes taken for a maple – but
ves alternate, the main veins not radiating
m the base. Often exotically bright red
umn colours. **Flowers** In showy creamy
ıds in late spring. **Fruit** The brown,
m berries taste of dates when over-ripe, and used locally to be brewed into an alcoholic drink (whence, it is assumed, 'The Chequers' as a pub name).
COMPARE Turkish Hazel (p.110); Cut-leaved Alder. In winter very like True Service (p.166). Wild Pear (p.177): twiggier; bark more closely square-cracked.

Service Tree of Fontainebleau
Sorbus latifolia

Probably originating as a hybrid of Wild Service, and confined in the wild to the Ile de France, but the most widely planted in the UK of this group of microspecies. Very occasional.
APPEARANCE **Shape** Broad, on sturdy horizontal branches; to 20 m. **Bark** Purple-grey; wide *papery-scaling* ridges. **Leaves** Glossy, with a *dull green-grey felt* beneath; broadly tapering at the base and *almost as broad as long*; small but consistent *triangular lobes tip the distant main veins.* **Fruit** Dull red-brown, 12 mm.
COMPARE Swedish Whitebeam (p.164): rounded lobes. Common Whitebeam (p.161): leaves less lobed, bright white beneath; smoother grey bark, and closer main veins.

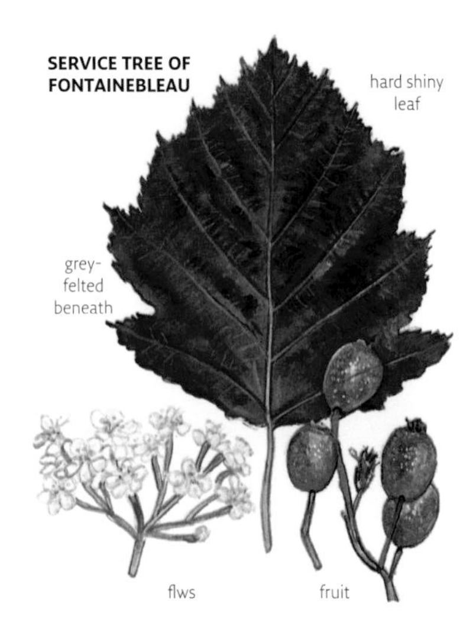

OTHER TREES Several comparable wild microspecies (planted only in a few botanic gardens) reach tree size in Britain and Ireland:

French Hales, *S. devoniensis*, locally frequent in Devon, E Cornwall and SE Ireland, has *rather narrower* and smaller leaves (9 × 6 cm).

Bristol Service, *S. bristoliensis* (Leigh Woods and Clifton Down near Bristol), is small-growing: leaves only 5 × 9 cm, whiter beneath, with *narrow-tapered bases* and veins crowded towards the rounded tip; fruit *bright orange*. Bigger orange-berried trees with narrow but longer, dully felted leaves, in a f town parks, are probably *S. decipiens* – who wild population, confined to the Burgberg Germany, is critically endangered.

Exmoor Service, *S. subcuneata* (cliffs on Exmoor), shares narrow leaves tapered at t base, but has browner berries.

S. × *vagensis* is a similar, infertile, broad-leaved hybrid of Common Whitebeam and Wild Service – wild in the Wye Valley and the Blean in Kent.

S. croceocarpa (*S.* 'Theophrasta') is an occasional park tree (naturalized on Anglese broad leaves *scarcely lobed* (rougher bark, darker crown, rounder leaves with fewer ma veins than Common Whitebeam); fruit mo like Service Tree of Fontainebleau's.

Swedish Whitebeam 
Sorbus intermedia

(*S. scandica*) Baltic region; long grown in the UK and sometimes seeding. Abundan a singularly tough street tree.

APPEARANCE Shape A lollipop dome of spreading, twisting branches; to 20 m. **Bar** Grey, remaining largely *smooth*. **Leaves** Ses Oak-shaped (but the round lobes are serra with *very narrow sinuses*), grey-woolly beneath; six to nine vein-pairs. **Flowers** In

ırticularly showy creamy-white heads. **ruit** Scarlet berries soon falling or eaten.
OMPARE Service Tree of Fontainebleau ›.163): leaves with triangular lobes. ommon Whitebeam (p.161): leaves, rarely ıore shallowly lobed, whiter beneath.
THER TREES *S. mougeotii* (Pyrenees to ustria) grows in a few collections: *more umerous small, close, rounded lobes*, often *ntoothed on inner side*; *10–12* vein-pairs.
S. anglica, from limestone areas of SW ngland, Wales, Co. Kerry, is a similar but ırubby microspecies (specimens in some otanic gardens have reached tree size). *S. ustriaca* (Austria to the Balkans) has leaves hiter beneath.
Arran Whitebeam, *S. arranensis* (N rran; bushy), has leaves with much deeper ısal lobes.

Bastard Service

Sorbus thuringiaca 'Fastigiata'

Wild (infertile) hybrids of Common Rowan and Common Whitebeam occur rarely; this erect variant (grown since the late eighteen century) is now quite a frequent street tree.

APPEARANCE Shape A neat *balloon of close, erect branches*; to 18 m. **Bark** Grey, smooth. **Leaves** Grey-woolly beneath, with *two to four free leaflets* on stronger shoots (to 14 leaflets in the rare clone 'Decurrens'). **Fruit** Red, in heads of 10–15.

OTHER TREES Finnish Whitebeam, *S. hybrida* (*S. fennica*; a group of stable hybrids of Common Rowan and Cliff Whitebeam), from S Scandinavia, makes *wide-spreading* trees in some gardens/streets, with shorter leaves. Arran Service, *S. pseudofennica*, is a native microspecies of this group, restricted to the N of the island. *S. meinichii* (W Norway) has four to six free leaflet-pairs.

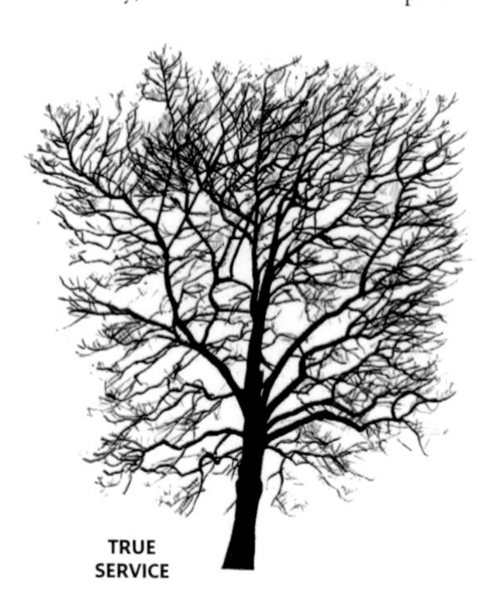

True Service
Sorbus domestica

(*Cormus domestica*) S Europe to the Caucasus; N Africa. The True Service's native status in Britain was disputed until the discovery of wizened trees perhaps 1000 years old on cliffs near Cardiff in 1984; one mysterious individual in Wyre Forest, ancient by 1678, had finally been burnt down by a tramp 184 years later. As a planted park tree it remains very occasional.

APPEARANCE Shape To 20 m; may be massively but gracefully domed. Dark yellowish crown of *drooping* leaves. **Bark** Black-brown; crisply *criss-cross, square-cracking ridges*. **Shoots, Buds** As Wild Service (p.163). **Leaves** As Commo Rowan (opposite) but slightly bigger; softly hairy beneath. **Flowers** Cream; big heads in late spring. **Fruit** *To 3 cm*, green brown; apple-shaped in some trees (f. *pomifera*), pear-shaped in others (f. *pyrifera*); edible when over-ripe. The variants are equally common and produc seedlings of either form.

ɔMPARE Black Walnut (p.96): ırk bark but much larger, nger-pointed leaflets. Wild ɛrvice: very similar in winter ɒark usually slightly paler and alier). Wild Pear (p.177): a ockier, twiggy tree in winter.

ommon Rowan ✪

ɔrbus aucuparia

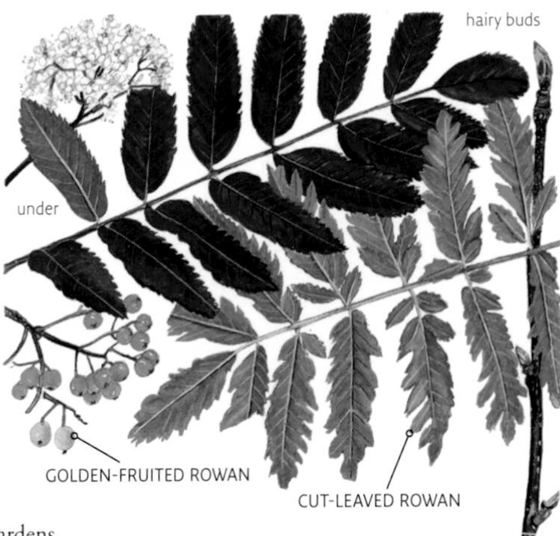

ɅIountain Ash) Europe, cluding Britain and Ireland; Africa; Asia Minor. Abundant ı light, acid soils, especially Highland Scotland. Planted ɛrywhere in streets, parks and gardens. ɔPEARANCE **Shape** Roughly and openly ire-shaped at first; branches always ɛanly ascending. Exceptionally to 25 m. **ark** Smooth and silvery-grey, horizontally ɛeaked with dark lenticels; scarcely rugged age. **Shoots** Purplish grey with hairs soon ed. **Buds** To 15 mm: *purple, non-sticky ıles are edged with long grey hairs* – like the ɒnic abdomen of some big spider. **Leaves** ɹpically with 15 rather *rectangular, often llowish* leaflets each 5 cm long and serrated within 1–2 cm of base; dense hairs neath are normally shed during summer. ıtumn colour orange – best in north. Saplings can have deeply toothed/ lobed leaflets (cf. Cut-leaved Rowan, below). **Flowers** Big, flat, creamy-white heads in late spring. **Fruit** Scarlet 1 cm berries soon eaten by birds.

COMPARE Other rowans: a highly ornamental group with berries of many colours and with a gamut of additional Asiatic species in big gardens. Most Common Rowans' rather oblong, dark-yellowish, matt leaves are distinctive. True Service (p.166): closest in foliage, but a large, craggy-barked tree. Others with

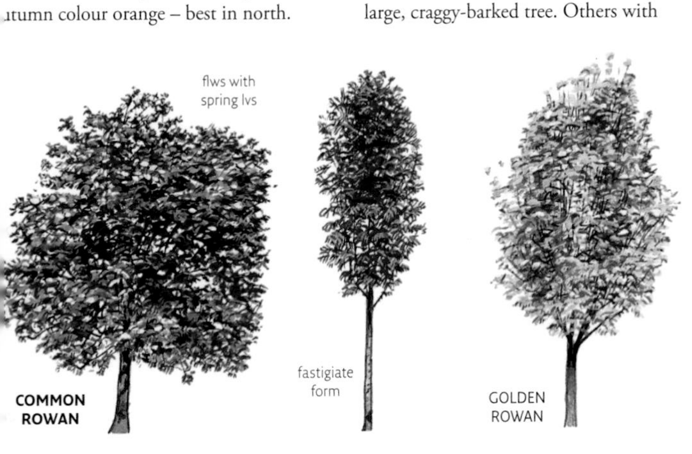

alternate, serrated, compound leaves (e.g. Black Walnut, p.96) have bigger foliage. False Acacia (p.198): untoothed leaflets. **VARIANTS** 'Sheerwater Seedling' is the most frequent of several *narrowly upright* selections – a shapely, conic, straight-trunked, very floriferous tree, with rather deeply toothed leaves.

'Beissneri' (1899) is quite frequent: a small, upright plant with yellowish green leaflets sometimes jaggedly lobed (cf. Cut-leaved Rowan) and remarkable bark, *dull pinkish-orange* and waxy, shining when wet but fading as algae colonize it. Autumn colour yellow; berries sweeter (cf. Edible Rowan).

Cut-leaved Rowan, 'Aspleniifolia', is rare: a tree with normal bark and habit but leaflets more downy beneath, and (even when adult) variously and *jaggedly lobed*/double-toothed.

Golden Rowan, 'Dirkenii', is upright; *bright yellow foliage* fades to soft green through summer. Rare.

Golden-fruited Rowan, 'Fructu Luteo' ('Xanthocarpa'; 1893), is rather occasional and has *golden-orange* fruits.

Japanese Rowan
Sorbus commixta

Korea, Japan, Sakhalin. *c.*1890. Rather occasional. (Sometimes labelled '*S. matsumarana*', '*S. discolor*' or '*S. serotina*'.) **APPEARANCE** Habit, bark and aspect much as Common Rowan (p.167). **Buds** *Crimson* (or green and red); variably *sticky* (cf. Sargent's Rowan, right). **Leaves** Rather deep glossy green; often held densely above the horizontal. Leaflets *longer-pointed* than Common Rowan's; hairless except sometime for brownish down under the midrib (extensive in var. *rufo-ferruginea*; cf. Ghose's Rowan); each to 7 cm. *Purple* and crimson autumn colours. **Flowers** A week after Common Rowan's; sweet-scented. **Fruit** Bright red, in great heads.
VARIANTS 'Embley' is a (slightly commone selection: smaller, *narrower, close-set,* 5 cm leaflets (cf. 'Joseph Rock', below); *scarlet* then deep crimson in autumn.

Sargent's Rowan 
Sorbus sargentiana

W Sichuan, China. 1910. Still rather rare. **APPEARANCE Shape** Often gauntly bushy; to 15 m. **Shoots** Stout, dark brown. **Buds** Big, *crimson and very sticky* (cf. Horse Chestnut's opposite buds p.216). **Leaves** Wi semicircular stipules at the base; 9–11 *long* (13 cm), pointed leaflets, softly downy benea and each with 20–25 pairs of sunken veins;

they flush dark brown and turn brilliant gol and crimson in autumn. **Fruit** Tiny (6 mm), but bright red and in heads of 200–500.

Joseph Rock's Rowan
Sorbus 'Joseph Rock'

(*S. rockii*; Rock 23657) A clone probably found wild in NW Yunnan (China); now occasional garden tree.

PPEARANCE Shape Upright, with
oreading branches, to 12 m; dainty. **Shoots**
oon hairless. **Leaves** *Small* (10–18 cm):
sually with 19 pointed, *closely set* 3–4 cm
aflets, fresh green and rather glossy. **Fruit**
'rimrose-yellow (white on the shaded side),
ı big heads – gorgeous among the crimson-
nd-purple autumn foliage.

'ilmorin's Rowan

orbus vilmorinii

S. vilmoriniana) Yunnan (China). 1904. A
articularly dainty, ferny-leaved rowan; very
ccasional.
PPEARANCE Shape Broad, on a trained
tem or bushy. **Shoots** With rusty hairs at
irst. **Leaves** Small (12 cm): up to 29 tiny,
ound-tipped leaflets, dark green above and
reyish beneath, wrinkled and almost hairless;
hey unfold pale brown and turn deep red in
utumn. **Fruit** 1 cm, *deep maroon-pink then
ipening to pink-flushed white.*

Hubei Rowan

Sorbus glabrescens

Yunnan, China. 1910. Occasional in parks and gardens; generally labelled *S. hupehensis* – an ally (from Hubei) scarcely grown in the UK.
APPEARANCE Shape Strongly though narrowly *domed*; to 17 m. **Bark** Grey-brown, developing *slightly corky ridges.* **Shoots** Sturdy, soon hairless. **Buds** Green then red. **Leaves** With 11–17 rather blunt, *glaucous* leaflets, silvery-grey beneath with tiny waxy warts but almost hairless; finely toothed only in the upper half. They unfold bronze, and turn lambent *pink* and red in autumn. **Flowers** In open, creamy-white puffs. **Fruit** *White, flushed pink.*

APPLES

Apples (about 30 species and several thousand hybrid cultivars, of which only the commoner ornamental forms are covered here) have berry-sized fruits, or 'apples'. Their sepals (which make the brown star at the bottom of eating apples) may be shed as they mature. Most trees flower together in late spring: Japanese Crab (p.172) and Purple Crab (p.175) are usually first, with Orchard Apple a week later.

THINGS TO LOOK FOR: APPLES

- Leaves: How broad? Are they hairy beneath?
- Flowers: Are they double? What colour are they?
- Fruit: What shape is it (rounded? conic? lemon-shaped?). How big is it? Are the sepals retained?

KEY SPECIES 

Orchard Apple (below): big, oblong leaves and 'apples'. **Japanese Crab** (p.172): slender leaves; 1 cm yellow fruit. **Hubei Crab** (p.173): longer, glossy leaves and scaly bark. **Purple Crab** (p.175): purplish leaves.

Orchard Apple
Malus pumila

(*M. domestica*). Fruiting apples have been bred over millennia from central Asian stock (not from Wild Crab); abundant in gardens and orchards and as 'wildings' in scrub and by roads or railways.

APPEARANCE Shape Low and wide, with many short, fruiting 'spur' shoots and vigorous, straight extensions; not spiny. **Bark** Shallowly scaly; greys, browns, some purples. **Shoots** Stout, grey; slightly *hairy*. **Buds** More or less *grey-woolly*. **Leaves** Thinly *woolly beneath* (cf. Pillar Apple, p.172) and on the stalk; *oblong*, large (to 12 cm), with small, irregular teeth; dark, rather matt and variably crumpled. **Flowers** From rich pink buds: white with soft pink shading. **Fruit** *At least 4 cm wide*; most are edible.

COMPARE *M. pumila* 'Elise Rathke' (p.176); Wild Crab (p.171); Bullace (p.191). Many 'wild' apple trees are clearly hybrids and can be tall-domed, to 18 m. Other crabs have narrower/more pointed, less downy leaves; Plum-leaved Crabs (p.174) are most similar.

THER TREES Danube Apple, *M. dasyphylla* ;E Europe; scarcely grown in the UK) is wild species with similarly downy foliage nd 4 cm apples.

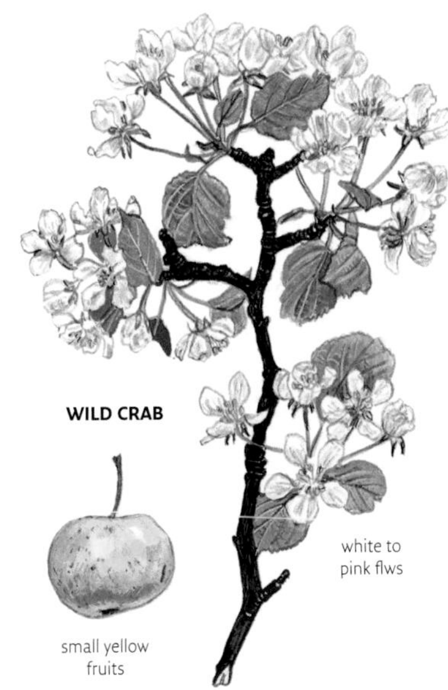

Vild Crab

1alus sylvestris

urope including Britain and Ireland; equent in old woods and hedges on heavier oils but until recently hardly ever planted. **PPEARANCE Shape** Irregular; to 17 m; ometimes spiny when young. **Bark** 'urplish brown; closely scaly ridges. **Shoots** *oon hairless* and glossy green/brown. **Buds** rown, downy only at their pointed tips. **eaves** Almost *hairless*, quite glossy; oval, to cm; rather folded; with small, rounded-riangular teeth. **Flowers** White, from pink uds. **Fruit** Crab-apples yellow-green, hard nd *very acid*, to 4 cm, dropping in winter nd often carpeting the ground.

COMPARE Orchard Apple (above). 'Wild' apples with pinker flowers, larger fruit or leaves downy beneath are hybrids. Snowy Mespil (p.179): leaves flatter and paler. Wild Plum: leaves more wrinkled. Wild Pear (p.177): similar winter shoots; blacker bark; leaves flatter and glossier.

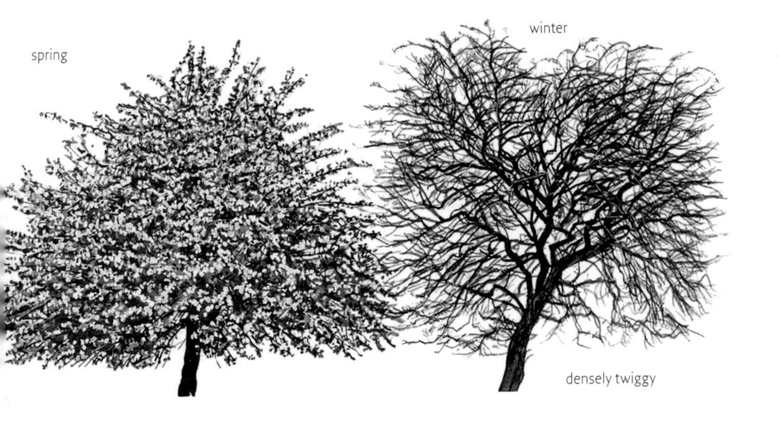

Japanese Crab ✪
Malus floribunda

Japanese gardens. 1862 (probably a long-cultivated hybrid). Abundant: a most dazzling flowering crab.

APPEARANCE Shape A low, particularly *tangled* dome of zig-zag branches, to 10 m; sparsely leaved by late summer. **Bark** Dull dark brown, *closely fissured* into knobbly oblongs. **Leaves** *Small, quite narrow*, pointed, 7 × 3 cm (cf. Myrobalan Plum, p.192), with the odd big lobe on strong growths (cf. Siebold's Crab); dull dark green above and finely downy beneath; stalk finely hairy. **Flowers** From rich carmine buds, ultimately white: very long-stalked, and so profuse they *totally smother the very early leaf-flush*. **Fruit** 1 cm: dull yellow; sepals shed; seldom conspicuous.
COMPARE Siberian Crab (p.174); *M. × scheideckeri* 'Excellenz Thiel' (p.176).

Pillar Apple
Malus tschonoskii

(Chonosuki's Crab; *Eriolobus tschonoskii*) Japan. 1897. Rare in the wild but now very frequent in parks and streets, principally for its vibrant autumn *golds and scarlets*; short-lived.

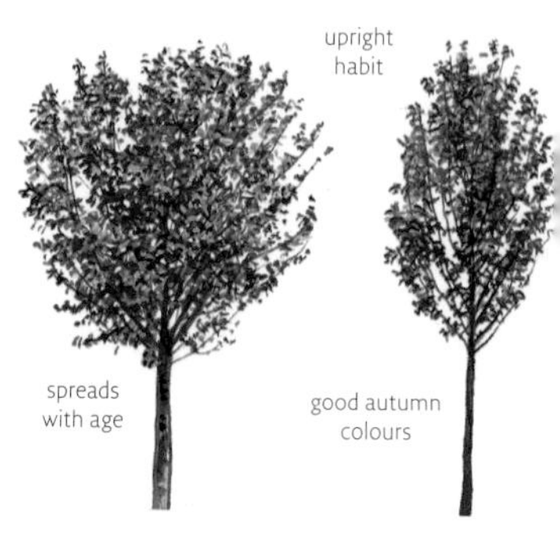

APPEARANCE Shape *Conic on steep branches*, to 17 m. **Bark** Grey; *smooth*. **Leaves** *Large*, to 12 × 8 cm, leathery; *grey-felted beneath*. **Flowers** White, among the *silvery* young leaves – restrained. **Fruit** 25 mm, yellowish with a purple cheek; sepals persistent. Seldom abundant.

OMPARE Grey Alder (p.105).
THER TREES Yunnan Crab, *M. yunnanensis* W China, 1908), is rare: *wide* crown; eener leaves with more distinct triangular bes (cf. Sweet Crab). Flowers showy, rongly azalea-scented; fruits smaller 2 mm), *deep red with white dots.*

Pratt's Crab, *M. prattii* (W China, 1904), as dotted fruit like Yunnan Crab's; its milarly large leaves, soon almost *hairless,* ay be narrower.

Iubei Crab ✪

Ialus hupehensis

M. theifera) Central and W China (with ndangered populations in Japan and aiwan). 1900. Quite frequent: a tree which ould be grown for its luxuriant foliage (used as tea in China) even if it did not flower spectacularly.

APPEARANCE Shape A *luxuriant* wide dome; to 17 m. **Bark** *Long, scaling, often spiralling plates,* with *orange/pink tints.* **Leaves** Typically *long* (to 10 × 5 cm), boat-shaped; flushing purplish then usually a *vivid, shiny green*; downy under the main veins; stalk woolly. **Flowers** White (pink in f. *rosea*) from pink buds, a few days after Japanese Crab's, petals *overlapping*; usually three styles. **Fruit** 1 cm, *dull* yellow/red; sepals shed.

COMPARE Siberian Crab (p.174): differences emphasized.

Siberian Crab
Malus baccata

NE Asia. By 1784. Occasional: streets and bigger gardens.

APPEARANCE Shape Lower: more twiggy and sprouty than Hubei Crab (opposite); to 15 m. **Bark** Usually browner and more closely cracking. **Leaves** Generally smaller, more matt; often soon *hairless*. The stalk is hairy only in the eastern var. *mandshurica*, which is common in the UK. (Its rare clone 'Jackii', from Seoul, has leaves as big and glossy as Hubei Crab's, but *blackish-green*.) **Flowers** Pure white from pink buds; petals *not overlapping*; usually five styles. **Fruit** *Bright* yellow/red, 1 cm; sepals shed and the stalk-end slightly sunken.

VARIANTS 'Lady Northcliffe', of uncertain parentage, has flowers darker pink in bud and dull, brownish fruits; rare.

COMPARE Hubei Crab (p.173).

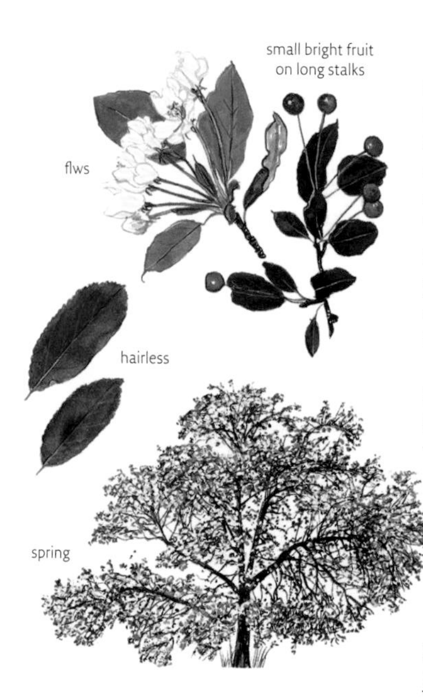

Plum-leaved Crab
Malus prunifolia

An apple of obscure, probably Chinese, origin; occasional in parks and gardens.

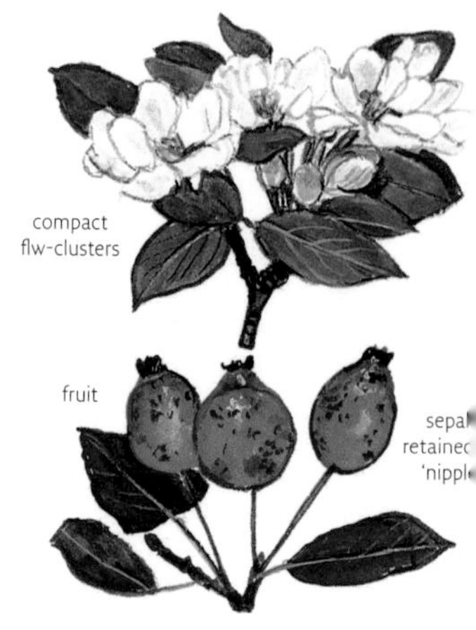

APPEARANCE Shape Quite dense and upright, to 10 m; dull, slightly greyish cast (like Snow Pear). **Bark** Grey-brown, often rather regularly fissured. **Leaves** Rather oblong, 6 cm; downy beneath. **Flowers** White, fragrant; less spectacular and delicate than Siberian Crab's. **Fruit** Profuse, greenish yellow to reddish; *lemon-shaped* (to 3 cm); sepals retained on top of the 'nipple'.

COMPARE Orchard Apple (p.170).

Cherry Crab
Malus × robusta

A group of hybrids of Siberian and Plum-leaved Crabs, very frequent in small gardens.

APPEARANCE Shape Spiky; frequently bushy. **Leaves** Quite large, coarse, oblong; downy beneath; early flushing. **Flowers** White from red buds (recalling Japanese Crab's (p.172), but duller and less

ıassed). **Fruit** Hanging *until next spring*: *berry-like, many with a star of sepals, some without*; golden in 'Yellow Siberian', crimson 'Red Siberian' and the popular 'Red entinel'.

OMPARE Siberian Crab (p.174): smaller uits; sepals always shed. Plum-leaved Crab).174): lemon-shaped fruit, always with epals.

THER TREES Some other comparable crabs ave spectacular fruit, often lasting into inter:

M. 'John Downie' (by 1891; frequent) as rather narrow leaves (6 × 3 cm), glossy reen on a hairy stalk; white flowers from ale pink buds in late spring. Fruits bright orange/red, to 4 cm (cf. 'Neville Copeman', p.176), *tapering*: they look glorious on the tree but put owners in a quandary as they taste delicious and make the best jelly.

M. × *zumi* 'Golden Hornet' (1949; abundant) has perhaps the richest fruit display of all. In summer, *gauntly spiky* and resembling Plum-leaved Crab (p.174); can be vigorous (to 9 m). White flowers from pink buds lead to great *wreaths of yellow, lemon-shaped 20 mm fruits*, with sepals, that rot brown on the tree.

Purple Crab ✪

Malus × purpurea

An abundant garden hybrid (by 1900).

APPEARANCE Shape Very untidy; often almost leafless by late summer. To 10 m. **Bark** Purplish grey, shallowly scaling/cracked. **Shoots** Black-purple. **Leaves** Unfolding glossy purple and fading to dark mauvish grey-green. To 6 × 3 cm; rarely three-lobed. **Flowers** Smouldering red-purple – quite dazzling in late sun – then mauve. **Fruit** Purple-red, round, 2 cm; sepals retained.

VARIANTS 'Lemoinei' (1922; probably now the commonest clone) has a denser crown, richer purplish green in summer, leaves to 10 cm (seldom lobed), and larger (5 cm) flowers with odd extra petals; fruits small, very dark purple. 'Aldenhamensis'

(1912) is upright and open, flowering later (with some extra petals); only the newest leaves are maroon in summer. Bark paler, shallowly scaly; fruits purple, *tangerine-shaped*; sepals shed. 'Eleyi' (by 1920; rare) has dark crimson flowers and *conic* purple 25 mm fruits; sepals often shed. A *much leafier* reddish green tree. 'Neville Copeman' (1952; occasional) is a poor, sparse plant, but its *round 4 cm brilliant orange-crimson fruits* (sepals retained) are briefly spectacular.

OTHER TREES *M.* 'Wisley Crab' (1924) has *large* (6 cm) purple-red fruits, with sepals.

M. 'Royalty' (1958), open and sprawling, keeps the *darkest foliage* (among which its purple flowers are rather lost); bark with coarse vertical ridges; fruits 20 mm, rich purple, sepals retained.

M. × *moerlandsii* 'Profusion' (Holland, by 1938; 'Lemoinei' × Siebold's Crab) has crimson flowers *soon fading to soft mauve* among purplish leaves which mature a shin[y] healthy, rather reddish green. Its *bright red* fruits are *small* and round (15 mm), their sepals shed. 'Liset' (a darker-flowered sister tree) is scarcer.

Weeping Purple Crab

Malus × gloriosa 'Oekonomierat Echtermeyer'

A garden hybrid (1914), frequent.

APPEARANCE Shape Shoots *hang to the ground* from low zig-zag limbs; the purplish young leaves fade to grey-green. **Flowers** Purplish mauve. **Fruit** Purple-red, 25 mm.

VARIANTS *M.* 'Royal Beauty' (1980) is a healthier, leafier improvement.

OTHER TREES Other weeping crabs include *M.* × *scheideckeri* 'Excellenz Thiel' (1909), effectively an untidily igloo-shaped Japanese Crab (p.172); the vigorous *M.* × *scheideckeri* 'Red Jade' (1935) with pink-flushed cup-shaped white flowers and cherry-sized *brilliant red* crabs (sepals shed) and *M. pumila* 'Elise Rathke' (1886), wide spreading, with large, yellow, *edible apples*.

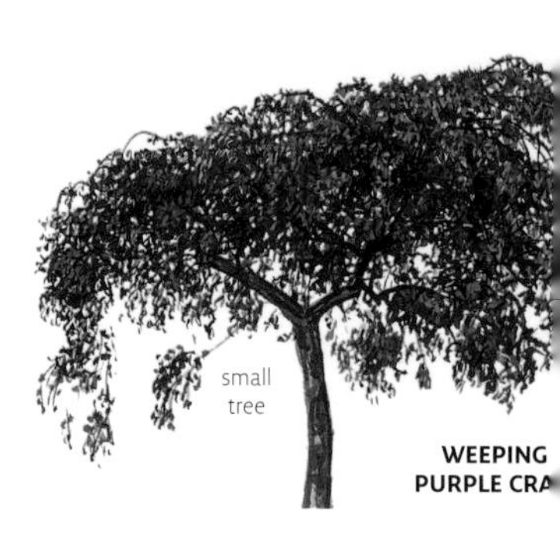

PEARS

Pears (about 30 species) have gritty fruits.

THINGS TO LOOK FOR: PEARS

- Bark: What is it like?
- Leaves: What shape are they? Are they downy beneath? Are they toothed?
- Fruit: How big is it? Are the sepals retained?

Common Pear ✪

Pyrus communis ssp. *sativa*

Fruiting pears (var. *culta*; many hundreds of clones) have been bred over millennia in Europe/W Asia from the wild species. Abundant in parks, gardens and, locally, old orchards.

APPEARANCE **Shape** Often gaunt and spiky; dense with fruiting spurs; very upright, strong shoots. Sometimes massively domed with age: clones like 'Pitmaston Duchess' reach 20 m and live for 300

years. **Bark** Black-brown, split – often as if by a knife – into *small knobbly oblongs*. **Shoots** Shiny brown, sometimes downy; rarely spiny. **Buds** Brown, pointed. **Leaves** Variable – often *heart-shaped* at the base. Rounded, narrowly oblong or slightly triangular, 3–8 cm long; *minutely round-toothed* (rarely untoothed). More leathery than apple leaves and glossier *blackish green* above; often hairless beneath but sometimes woolly (especially at first). **Flowers** Creamy-white heads overwhelm the tree a fortnight before apple-blossom. **Fruit** Pear-shaped in varying sizes, edible; sepals persistent.

COMPARE Wild Crab (p.171); Broad-leaved Cockspur Thorn (p.159).

VARIANTS Wild Pear, ssp. *communis* (*P. pyraster*) is the rare wild form; often tall-domed, with *spiny* young shoots and small rounded fruits (to 4cm; sepals retained), yellow-green and *hard even when dropping*.

'Beech Hill' is used as a younger street tree. *Strong, vertical shoots and younger branches* make a gauntly funnel-shaped but ultimately broad crown (to 14 m).
OTHER TREES Chanticleer Pear, *P. calleryana* 'Chanticleer', is an American selection of a Chinese species, much-planted in the late 20th century for its *neat spire-shape*, to 15 m so far. Bark *pale* grey-brown; rugged, largely vertical ridges. Flowers *very early* among silvery unfolding leaves which mature *brilliant green* and can persist until December.

Plymouth Pear
Pyrus cordata

(*P. communis* var. *cordata*) W France, Iberia and a few hedges around Plymouth and Truro – one of England's rarest wild trees (young plants are in some botanic gardens).
APPEARANCE Shape Shrubby and spiny. **Bark** Like Common Pear's. **Leaves** Small (4 cm), neat, rounded; seldom heart-shaped at the base. **Flowers** Among fresh-green young leaves; crimson stamens give a *pinkish* cast. **Fruit** *Like marbles*; sepals soon *shed*; edible when over-ripe.

small
rounded
lvs

Willow-leaved Pear
Pyrus salicifolia

Caucasus to N Iran. 1780. Very frequent: intensely silvery in spring, fading to iron-grey

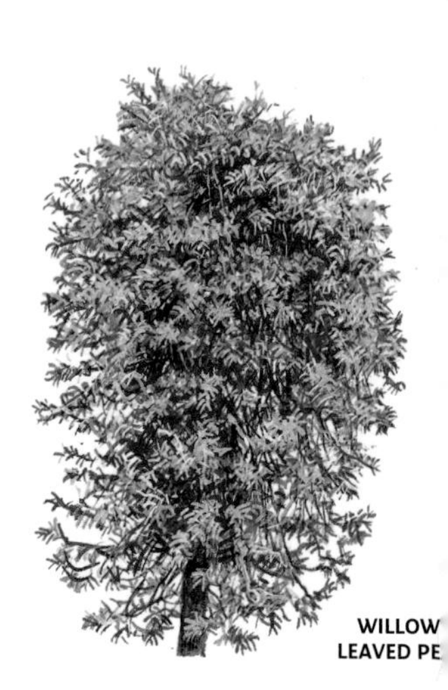

APPEARANCE Shape A low, tangled dome, to 13 m, the thin outer branches *weeping* except in shade. (Most grafts in gardens are probably the cultivar 'Pendula'.) **Bark** Black; ruggedly fissured. **Shoots, Buds** Silky-white. **Leaves** *Untoothed; silvery-hairy*, the upper side slowly turning shiny dark green; 4 × 1 cm to 9 × 2 cm. **Flowers** White, in well-spaced heads among the silvery young leaves. **Fruit** Pear-shaped, 3 cm; sepals persistent.
COMPARE Oleaster (p.227): silver *scales* under the leaves. Common Osier (p.92).

Snowy Mespil
Amelanchier lamarckii

(*A.* × *grandiflora*; Serviceberry; Juneberry) Abundant; commonly naturalized on sandy soils in SE England. Probably a stable hybrid of the E North American Shadblow (*A. laevis*, of which many specimens are grown but which has *hairless young leaves*): the nomenclature is much confused. Garden trees are also sometimes labelled *A. canadensis*, which is properly a North American suckering plant with shorter petals.

APPEARANCE Shape A low, twiggy but crisp dome, to 13 m, the trunk often very twisted. **Bark** Smooth, rather silvery-grey; neat shallow spiralling or criss-cross ridges can develop in old age. **Shoots** Slender, hairless. **Buds** Long-pointed and coppery – almost beech-like. **Leaves** *Flat but flimsy*, moss-green, unfolding coppery and with *silky hairs*, but soon hairless except on the stalk; oval, but with neatly toothed rather *parallel sides* halfway up. They glow rich orange and red in autumn, and individual jewel-like coloured leaves can be found throughout the season. **Flowers** With soft white, narrow, starry petals, in small heads above the young leaves in mid-spring: one of the most delicately beautiful of flowering trees. **Fruit** 9 mm berries ripen red then purple-black in mid-summer, and are soon eaten by birds.
COMPARE Wild Crab (p.171). Broad-leaved Cockspur Thorn (p.159).

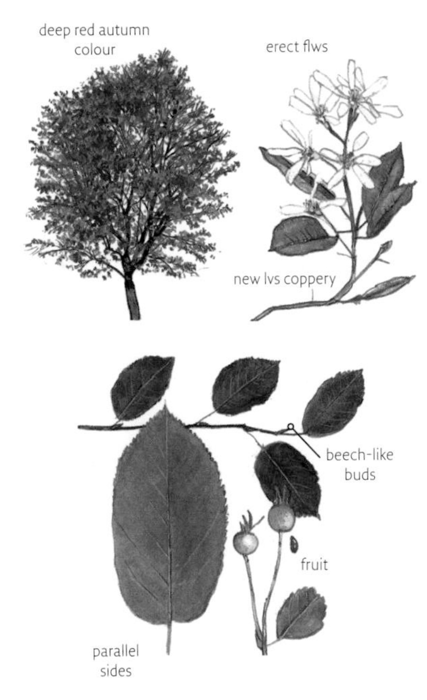

CHERRIES

Cherries (400 species) grow as trees or shrubs. The leaf-stalk normally has glands near the top (cf. willows, pp.87–93 and Hybrid Black Poplars, pp.82–6): these are nectaries, attracting ants to combat leaf-eating insects. Most are best learnt in flower.

THINGS TO LOOK FOR: CHERRIES

- Shape? Bark? What are they like?
- Leaves: How long are they? How hairy beneath? Are their teeth single or double (and whisker-tipped?)? Are the leaf-stalks hairy?
- Flowers: What colour(s) are they? How early? How many petals? How many styles, and of what shape? Are they among young leaves (of what colour?), or do they precede them? Are the sepals (what colour?) serrated?

KEY SPECIES 

Wild Cherry (opposite): big long leaves (with coarse teeth), and flowers in bunches: see 'Key Flowering Cherries', below.
Bird Cherry (p.193): big long leaves (with tiny teeth), and flowers in stiff tails.
Almond (p.190): long very narrow leaves.
Tibetan Cherry (p.183): showy bark.
Plum (p.191): smaller oval leaves.
Portugal Laurel (p.194): big long evergreen leaves.

KEY FLOWERING CHERRIES 

Wild Cherry: jaggedly toothed leaves (odd hairs below). **Cheal's Weeping Cherry** (p.185): a weeping dome. **Sargent's Cherry** (p.183): hairless leaves with sharp but scarcely whisker-tipped teeth. **Yoshino Cherry** (p.184): similar leaves on *downy* stalks.

Wild Cherry ✪
Prunus avium

(Gean; Mazzard; *Cerasus avium*) Europe, including Britain and Ireland; N Africa; W Asia. Frequent on rich/heavy soils, forming suckering stands. Abundantly planted everywhere: a favourite in amenity woodlands for its blossom and great early vigour.
APPEARANCE Shape Spire-shaped when growing fast, with annual branch-whorls. Old trees, *to 30 m*, spikily domed. **Bark** *Purplish* grey: horizontally peeling papery strips and rough lenticel-bands (craggily fissured in age; cf. Downy Birch, p.99). **Shoots** Brown, bloomed grey; hairless. **Buds** Long, rufous (cf. English Oak, p.119: cherry-buds are clustered *only on flowering spurs*). **Leaves** Big (to 11 × 6 cm), dull green, with *deep* (2–4 mm), *blunt, irregular but simple teeth*; finely hairy under the main veins. In autumn, gold and scarlet-pink. Stalk hairless, with typical two to five knobbly crimson glands near the top. **Flowers** Ivory-white, in posies (stalks don't branch), flooding the tree in mid-spring just as leaves unfold fawn-bronze. **Fruit** The cherries in mid-summer are sweet/bitter (not sour) but accessible only to birds.
COMPARE Sour Cherry and Schmitt's Cherry (p.182).
VARIANTS These include hundreds of relatively low, broad, fruiting cherries, abundant in cottage gardens and (locally, in drier climates) in orchards. (Others – Duke Cherries, *P.* × *gondouinii* – are hybrids with Sour Cherry, p.182.)

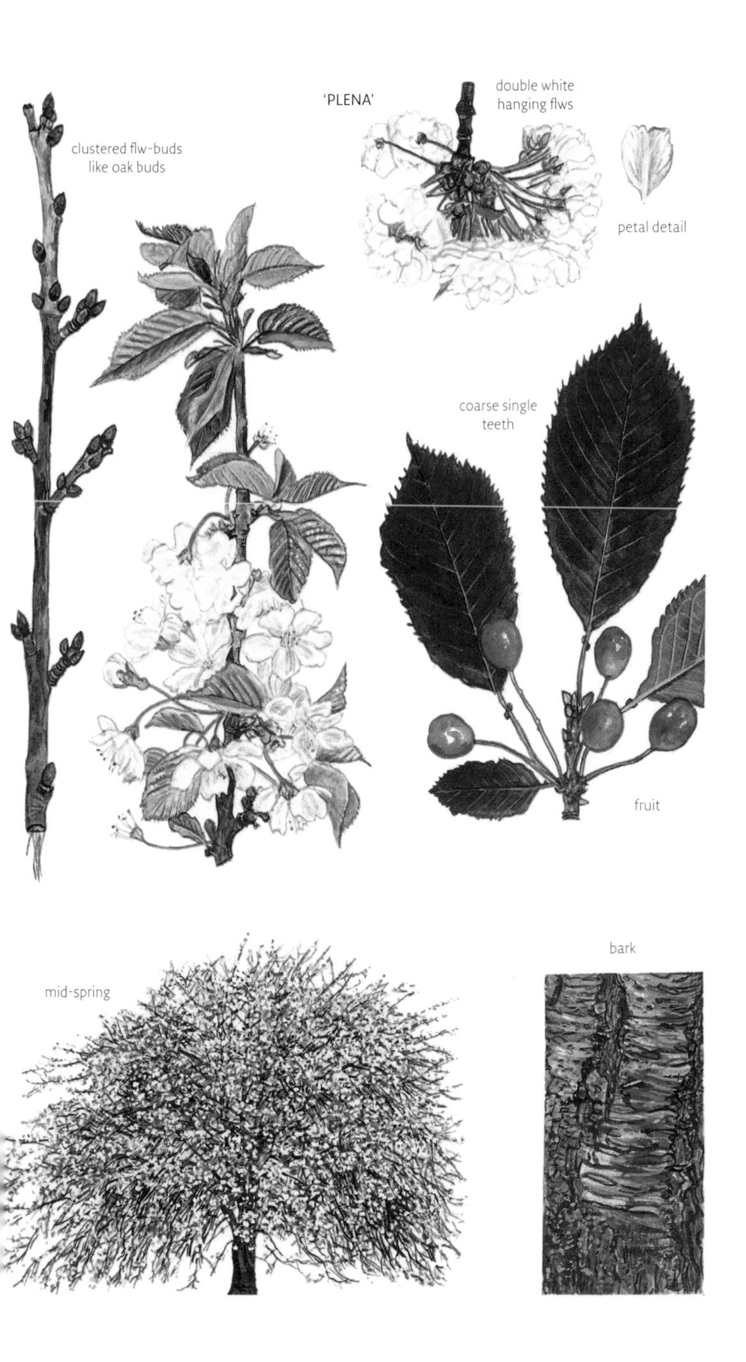
'PLENA'
double white hanging flws
clustered flw-buds like oak buds
petal detail
coarse single teeth
fruit
bark
mid-spring

Double Gean, 'Plena' (by 1700), is abundant: brilliant white, *double flowers, two weeks after the type's*, hang on long stalks under green young leaves. Bigger than other 'flowering cherries' (*to 20 m*); often identified by its broad but open, spiky-topped crown and elaborately *fluted*, rough-barked bole; its slender leaves tend to hang.

Weeping Gean, 'Pendula', is very rare: a parasol of hanging shoots from high, haphazard branches.

Sour Cherry

Prunus cerasus

(*Cerasus vulgaris*) Naturalized scarcely across Europe; probably not known wild.
APPEARANCE Shape Bushy; *abundantly suckering.* **Bark** Duller brown than Wild Cherry's. **Leaves** Smaller (to 8 × 4 cm; cf. Plum, p.191), a glossier sea-green above; usually *hairless; neat* teeth *finely rounded*, and rather *double*. **Fruit** Acid rather than bitter: Morello and Kentish Cherries are cultivars.
VARIANTS Rhex's Cherry, 'Rhexii', grown since the 16th century, is very occasional, carrying small (3 cm) double white pompom-like flowers at the end of spring among dark green young foliage. Leaves are smaller and darker than in other similar small 'flowering cherries'.

small lvs
SOUR CHERRY
flws
hairless
double flws
fruit stays sour
RHEX'S CHERRY

Schmitt's Cherry

Prunus × schmittii

Wild Cherry crossed with Greyleaf Cherry (*P. canescens*: a shrub in collections with narrow, downy leaves and a bark almost lik Tibetan Cherry's). 1923. Now a frequent street tree.

jagged teeth
flws
bark
big lvs

APPEARANCE Shape Unmistakable: *long, slender, erect, closely leafy branch* make a narrow, open balloon, pointe in youth. To 20 m; short-lived. **Bar** Finely peeling, *lustrous red-brown horizontal strips* between close roug lenticel-bands. **Leaves** Most like Wild Cherry's but shorter; coarsely toothed **Flowers** Pale pink; small and fleeting among olive unfolding leaves.

OMPARE Sargent's Cherry 'Fastigiata' (.184): more open; bark dull brown.
THER TREES Dawyck Cherry, *P.* × *awyckensis* (1907; probably another hybrid f Greyleaf Cherry, to 8 m in some big ardens), shares the bright bark; its sparse, *untly drooping shoots* are strung with eeper pink, showier flowers.

ibetan Cherry ✪
runus serrula

/ China. 1908. Occasional in gardens nd newer street-plantings – entirely for its nique bark; regenerating freely.
PPEARANCE Shape A finely twiggy ome of hanging, greyish leaves on rather iff branches; to 15 m. **Bark** *Crimson, tin-smooth* between rough brown bands; redding shaggily with age; ultimately acked, dull and sprouty. Conscientious wners scrub their trees with toothbrushes. **hoots** Shortly downy. **Buds** Long, slender, hestnut. **Leaves** *Narrow*, to 11 × 3 cm; sually silky-hairy under the veins. **Flowers** ellow-white, tiny and fleeting among the liage. **Fruit** Cherries red, oval, 15 mm long.
OMPARE Schmitt's Cherry (opposite). Dull-barked old trees can even recall lmond (p.190).

Sargent's Cherry ✪
Prunus sargentii

Mountains of Korea, Sakhalin and N Japan (where it makes a huge timber tree). 1893. Very frequent.

APPEARANCE Shape Rounded; shapely, with light, ascending branches; to 15 m; *densely clad* in *dark yellowish*, rather matt, hanging leaves. **Bark** Purple-brown, with orange lenticel-bands: a *browner* cast than

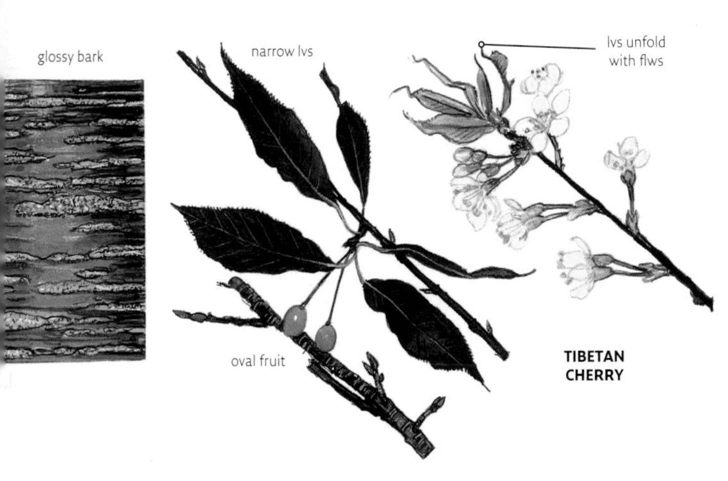

other 'flowering cherries'; more rugged in age. (Garden trees are often grafted on a stem of Wild Cherry.) **Leaves** Hairless and on hairless stalks, the sharp teeth scarcely double *and scarcely whisker-tipped*; rather broadly oblong (to 14 × 8 cm) and generally *convex* (the tip down-curving); whitish beneath. Autumn colours *early and scarlet*. **Flowers** Brilliant mauve-pink, single but quite large (4 cm), *daintily half-filling the crown* a week before the first Japanese Cherry, on non-branching stalks among *ruby-red baby leaves which are still dark bronze as the last petals drop*.
COMPARE Yoshino Cherry (right): hairy leaf-stalks. Sour Cherry (p.182): similarly hairless but shorter leaves.
VARIANTS 'Fastigiata' is occasional: vigorous, *gaunt, erect stems*, to 16 m. 'Rancho' (1961; rare?) is more compactly columnar.
OTHER TREES Hill Cherry, *P. jamasakura* (*P. serrulata* var. *spontanea*), from Japan, is undeservedly scarce. Bark pewter-grey; habit dainty, with slenderer leaves whiter beneath; flowers often white, on branching stalks; vivid clouds overtaken by usually red young leaves.

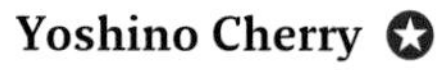
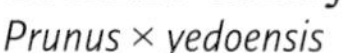

Yoshino Cherry ✪

Prunus × yedoensis

Now the most abundant garden cherry in Japan. *c.*1910; frequent in the UK.

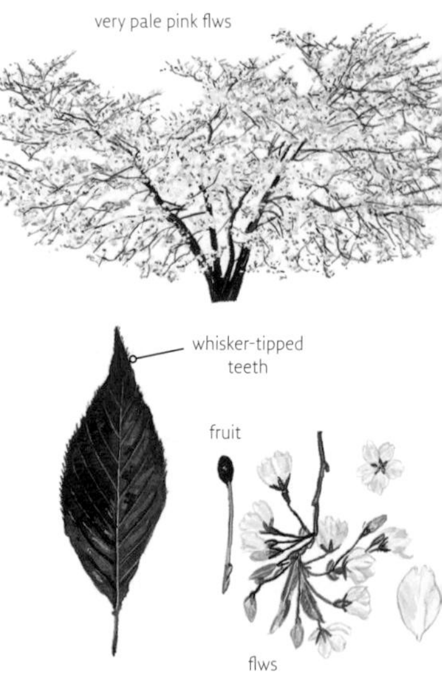

APPEARANCE Shape Usually low and dense (slightly weeping at the edges), on *heavier, more zig-zag limbs* than other 'flowering cherries'; to 15 m. **Bark** Pewter-grey, smooth; brown lenticel-bands. **Leave** Large, dark, olive-green, rather hanging;

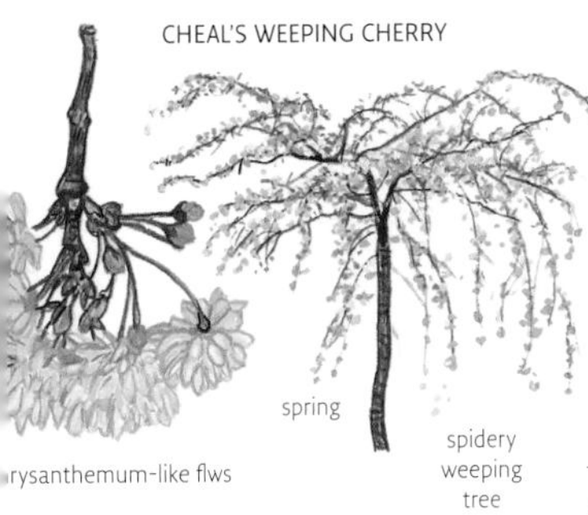

hisker-toothed; downy under the main
:ins *and on the stalk.* **Flowers** Pale blush,
1 unbranched stalks, slightly scented and
ensely wreathing the graceful shoots *ahead
its leaves.*
OMPARE Other 'flowering cherries' with
airs largely confined to the leaf-stalks:
Winter Cherry (below).
ARIANTS The frequent 'Shidare Yoshino'
Perpendens'; 'Pendula') makes a *neat,
ense igloo* of hanging shoots (*P.* 'Hilling's
Weeping' is a *white* variant); 'Ivensii'
by 1929; also white) *weeps from widely
rching limbs.*
THER TREES Weeping Cherries, whose
ranches arch to the ground from a high
raft, also include the following.
Cheal's Weeping Cherry, *P.* 'Kiku-shidare-zakura', is an abundant Japanese Cherry. Mauve-pink, very double flowers wreathe the sparse, gaunt branches in solid pompom clusters, among olive-green young foliage. The slender leaves (scarcely whisker-toothed) mature glossy and puckered.

Winter Cherry

Prunus × subhirtella 'Autumnalis'

('Jugatsu-zakura') Japanese gardens. *c.*1900. Very frequent.
APPEARANCE Shape Very twiggy and untidy; spreading, to 9 m. *Bare lengths of thin, twisting shoots* with small leaves at the tips; rather *pale moss-green foliage.* **Bark** Grey-brown, smooth; brown lenticel-bands. **Leaves** Small (7 × 4 cm), often down-curling near the tip; doubly toothed though not so deeply as Fuji Cherry's; hairy under the main veins and on their stalks. **Flowers** Small, semi-double, *almost white* (crimson stamens make them look pink); opening little by little *from October* until mid-spring when there is a final, quiet flourish at the branch-tips, among pale green young leaves.
VARIANTS 'Autumnalis Rosea' has *soft pink* flowers and is slightly less common.

'Japanese Cherries' (*Sato-zakura*) were bred in Japan (and China) over many centuries from wild species (though 'Pink Perfection' and the still-rare 'John Baggeson' are seedlings of 'Kanzan' raised in England); sometimes corralled as cultivars of *Prunus serrulata*. All are short-lived in the UK.

APPEARANCE Shape Generally low; customarily grafted at head-height on a stem of Wild Cherry, which outgrows the branches above it like a pillar box. **Bark** Typically pewter-grey, with brown lenticel-bands. **Leaves** Large, sparse; hairless and on hairless stalks; with more or less simple but strongly whisker-tipped teeth; autumn colours rich ambers and crimson-pinks. **Flowers** On variably long, branching stalks (cf. only Wild Hill Cherry, p.184): larger and generally later than those of other 'flowering cherries', often double, and opening when the (often red) young leaves are half-expanded.

Prunus 'Kanzan' (c.1913) ✪ is the most abundant and has given its whole clan a bad name. (The Japanese have the sense to use it sparingly). Briefly gorgeous as magenta buds swell under red baby leaves; less so as the 23–28 petals fade to hard pink and the leaves expand olive-green. Funnel-shaped when young; vigorous (to 14 m); heavy limbs soon curve untidily to the horizontal. Leaves large, very sparse; tomato-red growth-tips through early summer.

'KANZAN'

'KANZAN'

P. 'Pink Perfection' (Surrey, 1935; frequent has bright but mottled pink double flowers timed with 'Kanzan' but hanging among *quickly green* young leaves. In summer a rather *globular* tree with *drooping outer shoo* and *short*, dark leaves.

'PINK PERFECTION'

P. 'Amanogawa' (Lombardy Poplar Cherry) is planted abundantly for its *vertical* shape rather than its flowers, which (opening wit those of 'Kanzan') are also in erect posies,

'AMANOGAWA'

.e 6–15 pale pink petals a poor contrast to
:onzy-green young leaves. Old trees are as
·oad as tall but still have twisting, stiffly
›right twigs.

·runus 'Ichiyo' is a magnificent but only
cally frequent Japanese Cherry. Flowers 10
ıys before 'Kanzan', among young leaves
hich are soon *fresh green*; pink in bud
ding to *creamy* blush: a flat, neat ring of
5–22 petals *like a ballerina's tutu* surrounds

:een/red eyes (with 1–2 styles) that stare
: you from all over the tree. In summer, a
igorous cherry (often on its own stem) with
sing then spreading limbs, and relatively
nall, dull, narrow leaves; its bark develops
ugged vertical fissures, orange at the base.

· 'Ukon', very frequent, has 9–15 petals
'ushed greenish yellow, a week later; young
aves coffee-brown. Summer foliage as
Tai Haku' (habit – a straggling funnel, as
Kanzan' – and paler growing tips will often
ifferentiate).

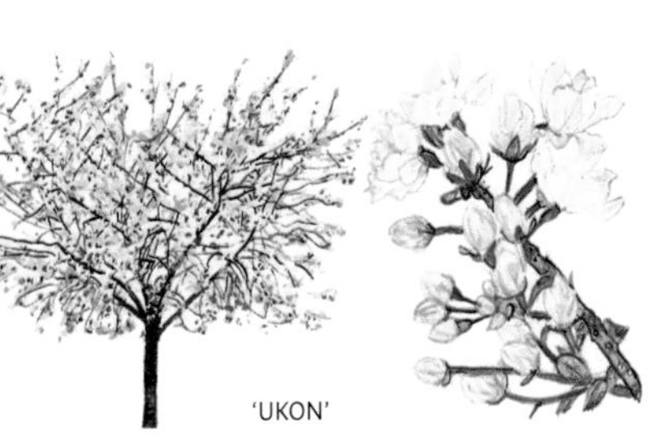

'UKON'

'SHIROTAE'

P. 'Shirotae' ('Mount Fuji'; *c.*1905) is very frequent: the first 'Japanese Cherry' to flower: dazzling white, honey-scented blossoms with 6–11 petals hanging under green baby leaves. Unique in summer: strong *very horizontal* branches (taller and gaunt in shade) carry bright rich green leaves, their teeth with *extra-long* crimped whisker tips.

P. 'Shogetsu' ('Shimidsu Zakura'; 'Longipes') is frequent: pink buds open a week after 'Kanzan' into pure white flowers with 20–28 petals which hang under yellow-green young leaves. Small, rounded, slightly weeping crown, often cankered and dying back.

'SHOGETSU'

P. 'Shirofugen', the last to flower, is very frequent and unsurpassed: almost white, double flowers hang from pink buds under *maroon* young

'SHIROFUGEN'

leaves, then *fade soft pink* as the leaves turn green. In summer as 'Fugenzo' – a vigorous, spreading cherry: medium-sized, dark, often oblong leaves, the whisker-tips of the teeth generally appressed.

Hybrid Flowering Cherries

Prunus 'Spire'

(*P.* × *hillieri* 'Spire') A 1930s hybrid of Fuji Cherry; locally abundant as a street tree.
APPEARANCE Shape Erect thin limbs usually from a very disproportionate Wild Cherry pedestal; densely leafy, compact but irregular; finally as broad as tall (to 12 m). Leaves To 8 × 5 cm, almost round, with a sudden long point (cf. 'Umineko'), and deep, rounded double teeth; slightly downy beneath; scarlet in autumn. **Flowers** 4 cm; blush-pink clouds, among red baby leaves; a week after 'Pandora'.

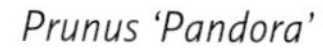

Prunus 'Pandora'

A complex garden hybrid (by 1940); quite frequent.
APPEARANCE Shape Funnel-shaped, then broad; quite graceful. **Bark** Dull grey-purple; rugged with age. **Leaves** Small (7 × 3 cm), glossy, blackish, neatly and sharply double-toothed; almost hairless; they hang rather sparsely by late summer. **Flowers** 4 cm: 5 petals, pale pink but dark near their margins: early and long-lasting, overtaken at last by bronze young leaves.
COMPARE Other small-leaved forms – 'Okame' (below); Tibetan and Sour Cherr(p.182–3).

'PANDORA'

'SPIRE'

Prunus 'Umineko'

(Seagull Cherry) Fuji Cherry crossed with Oshima Cherry (*P. speciosa*). 1948. Frequent.
APPEARANCE Shape Dense; neatly upright at first (like a seagull with raised wings); branches arch over at 7 m. Dark, vivid green; clear orange in autumn. **Leaves** Downy-stalked; almost round with a sudden long point, to 9 × 6 cm; deep, neat, double teeth: cf. 'Spire' (above: with blunt toothing) and Fuji Cherry (small, narrower leaves). **Flowers** Similar to Wild Cherry's (p.180) but snowier and earlier; slowly overtaken by green young leaves. Understated but very lovely.

Prunus 'Okame'

Fuji Cherry crossed with the red-flowered Bell Cherry (*P. campanulata*). 1947. Rather occasional.
APPEARANCE Shape Usually dwarfishly globular; stiffly twiggy and untidy. **Leaves** Small, dark, narrow (6 × 3cm), with odd hairs above at first; jaggedly and irregularly toothed (even lobed on saplings). **Flowers** Small, dark, smoky magenta-pink clouds from almost crimson buds, a month before the leaves.

Prunus 'Accolade'

Sargent's Cherry × *P.* × *subhirtella*. 1952. Frequent.
APPEARANCE Shape Spreading: very untidy, tangled, hanging, twisting bare shoots, like Winter Cherry's (p.185). **Leaves** Bigger (to 11 × 5 cm) and darker; hairy-stalked. **Flowers** The tree's redeeming feature: 4 cm; about 12 brilliant then pale pink petals (cf. Winter Cherry): superabundant strings and clusters, well before the soft-green leaves and often starting late in winter.

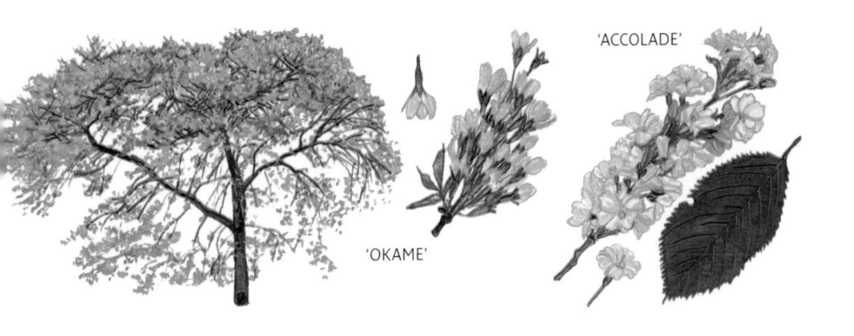

Almond ✪ ☠
Prunus dulcis

(*P. communis*; *P. amygdalus*; *Amygdalus dulcis*) Mediterranean regions; long grown in the UK. Flowering selections are frequent in small gardens.

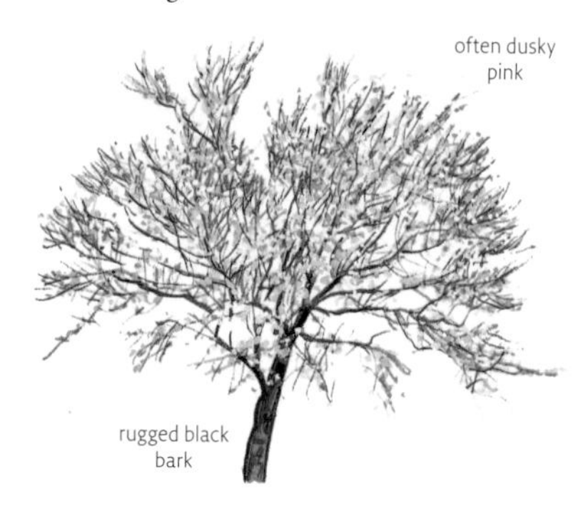

APPEARANCE **Shape** Gauntly globular, open and rather erect, to 11 m; wild trees can be spiny. **Bark** *Blackish; soon finely but ruggedly cracked.* **Shoots** Like Plum's (opposite), but with slightly downy buds. **Leaves** Willow-like, to 12 cm; dark and glossy but crumpled (often folded up at the base); hairless; finely serrated; on stalks *to 25 mm*, which carry the typical cherry glands. **Flowers** Big (to 5 cm), mauve-pink; sparsely single or paired on very short stalks; well before the leaves. **Fru** Like small greenish peaches, to 5 cm; finall dark brown. Chemicals in the almonds of unselected trees (var. *amara*) can release dangerous quantities of cyanide – 50 may be fatal.

COMPARE Black Cherry (p.194); Bay Willow (p.89).

OTHER SPECIES Peach, *P. persica*, from N China, has long been grown here buts need plenty of warm sunshine to ripen the sweet 8 cm peaches. The leaves have shorter stalk (12 mm); flowers smaller, paler and later.

Apricot
Prunus armeniaca

(*Armeniaca vulgaris*) N China; cultivated and naturalized in S Europe but rare outdoors in the UK.

APPEARANCE Shape Gauntly globular, on twisting branches. Bark Soon rugged; paler grey than Almond's. Shoots (Like the plums') *lacking a big end bud*. Leaves *Rounded* with a sudden twisted tip (particularly broad in var. *ansu* from E Asia); round-toothed; hairless or with tufts under the vein-joints. Flowers Rather small (25 mm) and early; white/pink-flushed. Fruit Apricots of unselected trees are only 3 cm wide.

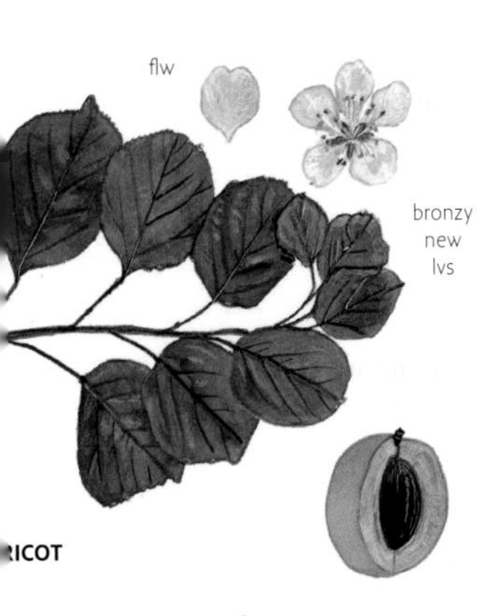

lum
unus domestica

ong grown in the UK; probably a aucasian hybrid of Blackthorn and yrobalan Plum. Abundant: gardens, some chards; suckering and naturalizing.
PPEARANCE Shape Low, dense; to 10 m; *ot* spiny. **Bark** Purple; finely oughened then widely ssured. **Shoots** Red-purple, oon *hairless*. **Buds** Longer nd sharper than the other ums' on this page (all *lack big end-bud* and can have vo to three side-buds at each af). **Leaves** Wrinkled (cf. oat Willow, p.91); to 8 cm, oadest above halfway; downy nder the veins; stalks with the typical cherry glands. **Flowers** Off-white, in mid-spring. **Fruit** At least 3 cm long.
COMPARE Sour Cherry (p.182); Orchard Apple (p.170).
OTHER TREES Bullace, *P. insititia* (*P. domestica* var. *insititia*), abundantly naturalized in hedges, is often bushy. Fine purple-grey shoots, *downy for a year*, are sometimes spiny; leaves smaller (but broader than Blackthorn's), with fine *down on both sides*. Fruits round, 2–3 cm; purple ('Black Bullace') or yellow ('Shepherd's Bullace'); *sweet*. Damsons (sweeter oval purple plums), Mirabelle Plums (round yellow plums) and Greengages (yellow-green oval plums) are rarely naturalized selections.

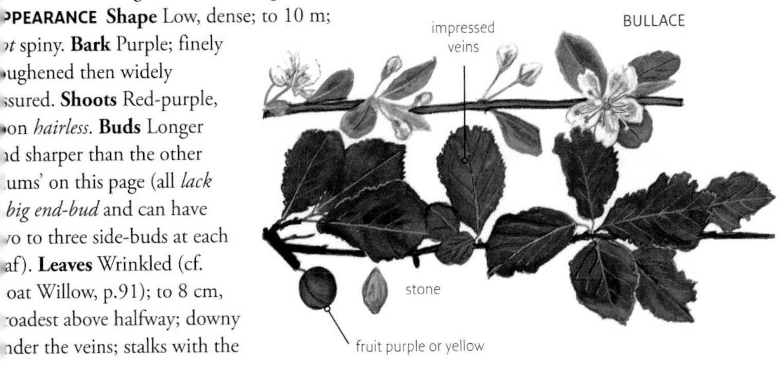

Blackthorn
Prunus spinosa

(Sloe) Europe, including Britain and Ireland; N Asia. Dominant in scrub; hedges on heavier soils.

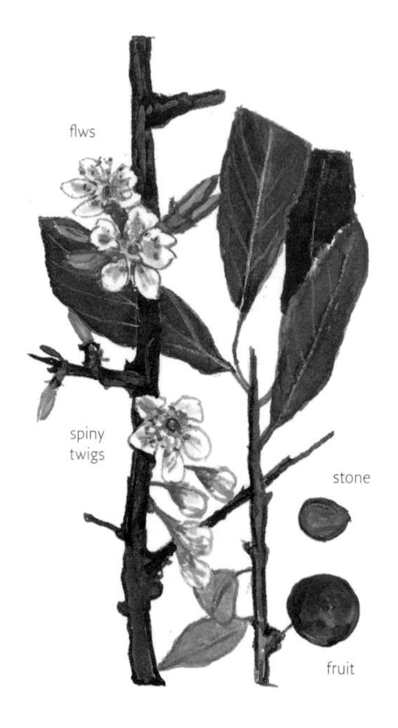

APPEARANCE Shape Generally a suckering *bush*. **Bark** Purple-black, finely roughened. **Shoots** Finely downy then *almost smooth by winter; gleaming purple* or grey-bloomed (green in shade); *many* side-shoots end in vicious spines. **Buds** 1–2 mm. **Leaves** Small, *slender* (to 5 × 2 cm), broadest above halfway; wrinkled; downy when young. (Their stalks often *lack* glands.) **Flowers** Off-white, in mid-spring but before the leaves. Sloes purple then black, 15 mm, still intensely *sour* by winter.

VARIANTS 'Purpurea', has purple leaves (smaller and paler than Pissard's Plum's) and pink flowers.'Plena' has double white flowers. Both are very rare.

Myrobalan Plum
Prunus cerasifera

(Cherry Plum) Long grown in the UK; abundantly naturalized. (Wild trees – Balkans to central Asia – may be distinguished as *P. divaricata*.)

APPEARANCE Shape Larger than other plums: *to 15 m*. **Bark** Dark *grey*; widely fissured with age. **Shoots** Soon *hairless; green*. **Leaves** Fresh green; slender, often broadest *below halfway*; downy under veins. **Flowers** Whiter than Blackthorn's and *much earlier* (weather-dependent). **Fruit** Sweet, 4 cm, round, red/gold plums, ripe *late in summer*.

VARIANTS Pissard's Plum, 'Pissardii' ('Atropurpurea'; Iran, 1880), is abundant: untidily upswept, with very pale pink flowers, shiny purple shoots and *heavy purple foliage*. 'Nigra', replacing it as a more graceful plant (scarcely separable in summer) has *deep pink flowers* a few days later.

'Hessei', low and twiggy, has blotchy *bronz red* leaves and white flowers; rare but muc prettier.

'Lindsayae' (Iran, 1937) is rare: pink flowe

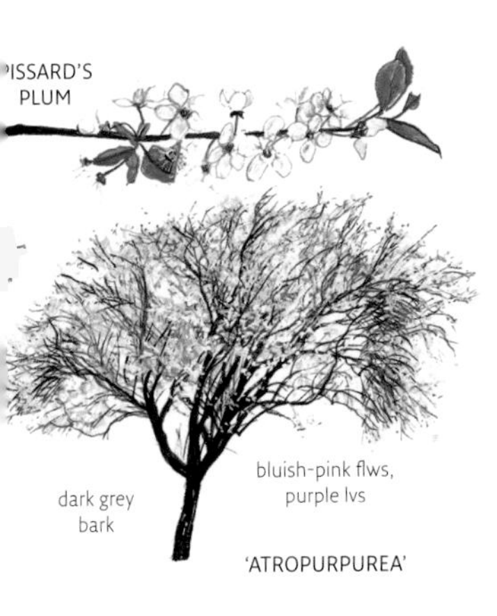

THER TREES Double Cherry Plum, *P.* × *lireana* 'Moseri' (Pissard's Plum × Japanese pricot, 1895), is quite frequent: lower, vith broader red-purple leaves and *double* cm flowers, rich pink.

ird Cherry ✪

runus padus

Hawkberry; Hagberry; *Padus racemosa*) N urasia, including hilly, limestone parts of reland and of Britain S to Norfolk; locally requent. The type has seldom been grown n gardens, but suckers from grafts may aturalize.

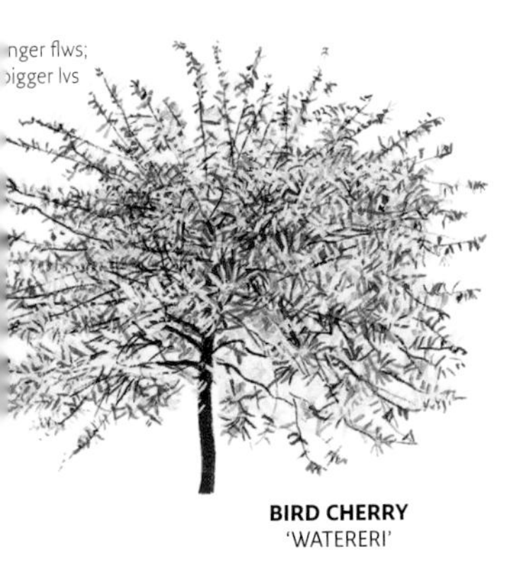

BIRD CHERRY
'WATERERI'

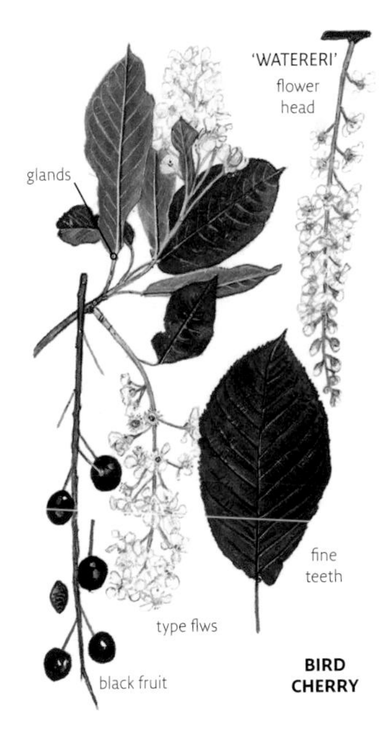

BIRD CHERRY

APPEARANCE Shape Conic and whorled, then openly domed (to 14 m): straight, rising limbs and fine, sweeping twigs. **Bark** Dull grey; very finely roughened but *never fissured/peeling*. **Shoots** Soon hairless; dark, dull green-brown with pale lenticels and sharp 1 cm buds – like Aspen's (p.79), but lacking perpendicular side-shoots. **Leaves** Dull green, and hairless except for tufts under the vein-joints; much *finer, sharper serrations* (1 mm deep) than Wild Cherry's (p.180). The hairless stalks have the typical cherry glands. **Flowers** In well-separated *stiff tails 8–15 cm long* from the shoot-tips in late spring. **Fruit** 8 mm, bitter, black cherries.

COMPARE Black Cherry (p.194).

VARIANTS 'Watereri' (1914), the common garden clone until recently, *grows to 25 m*. Very open and untidy with curving branches; larger (15 cm) leaves, always tufted beneath; much *longer* (to 20 cm), horizontally spreading flower-heads.

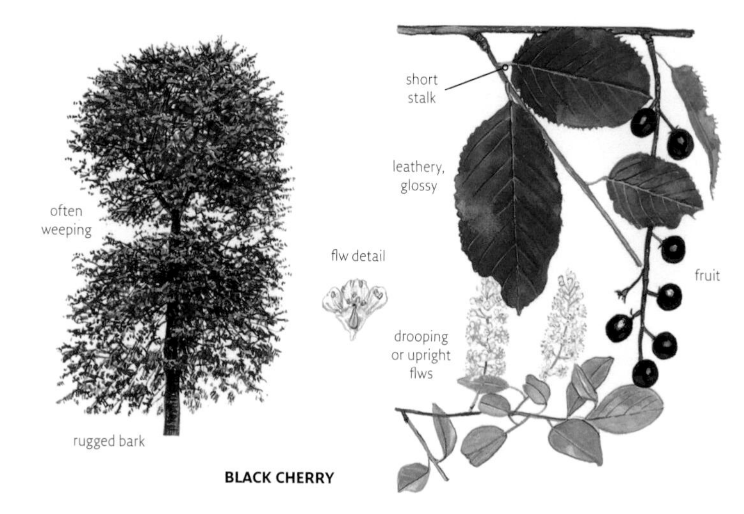

'Albertii' is occasional as a younger tree: neatly and narrowly ascending branches make a cone/egg-shape, at least in youth. Flowers a few days earlier, in short but dense, broad heads.

'Colorata' is occasional as a gawky young plant, strikingly beautiful in flower when *soft pink* tails stand among pale *bronze-purple* young leaves. In summer the shoots are purple and the leaves dull grey-green, with a purplish cast beneath.

Black Cherry

Prunus serotina

(Rum Cherry) Nova Scotia to Florida and Arizona. 1629. Very occasional in gardens, but sometimes well naturalized on sandy soils.

APPEARANCE Shape A somewhat shapeless but quite dense, evergreen-looking tree (late into leaf and gold late in autumn); outer branches sometimes very weeping. To 20 m. **Bark** Black-brown; soon *fissured and peeling harshly*, and rugged in age. **Shoots** Slender; glossy red-brown, with 4 mm yellowish buds, usually blunt. **Leaves** To 12 × 4 cm (but some much smaller); rather leathery; dark and glossy above and smooth beneath but with *dense orange/white hairs bristling from the prominent lower midrib*; fine *incurving* teeth. Stalk to 15 mm only, carrying the typical cherry glands. **Flowers** In tails like Bird Cherry's, in early summer. **Fruit** 8 mm crimson then purple-black cherries, which were used to flavour rum and brandy.

COMPARE Portugal Laurel (below); Almond (p.190).

Portugal Laurel

Prunus lusitanica

Iberia. By 1648. Abundant in hedges, shrubberies and game-coverts; thoroughly naturalized in the UK in woods on heavy soils, where it spreads by layering and may smother the native vegetation. Ironically it is threatened in the wild by an increasingly dry climate.

APPEARANCE Shape Bushy or on a sturdy, irregular trunk, to 18 m; densely and handsomely evergreen. **Bark** Dark grey; finely roughened. **Leaves** Very glossy; thinly leathery, flat and hairless; finely serrated and broadest below halfway; to 12 cm. **Flowers** In narrow, arching, fragrant tails to 25 cm

ng, turning the tree creamy-white in early mmer. **Fruit** 1 cm bitter purple 'cherries'.
OMPARE Cherry Laurel (below).
hododendron ponticum, which misbehaves the UK in the same ways, is sometimes nfused, but has scaly reddish bark and ntoothed leaves.
ARIANTS 'Variegata' (brightly but arrowly white-margined leaves, flushing ellow) is rare.

Cherry Laurel 

Prunus laurocerasus

Asia Minor to Iran; Bulgaria; Serbia. 1576. An abundant evergreen in hedges, shrubberies and old game-coverts, where on moist heavy soils it colonizes woodlands even more aggressively than Portugal Laurel (above). Used sparingly in herbal medicine ('Cherry Laurel water'), but chemicals in the leaves can release dangerous quantities of cyanide: clipping and burning it are best avoided.
APPEARANCE **Shape** Bushy (exceptionally to 18 m): sprawling, layering stems and branches gauntly clothed with big (to 20 cm) leaves. **Bark** Blackish; very finely roughened. **Leaves** Bright deep green; shiny, thickly leathery and hairless; broadest above the middle and slightly *convex*, with indented veins and *distant* tiny teeth. **Flowers** In erect white tails to 12 cm long in mid-spring. **Fruit** 15 mm blackish 'cherries' (toxic in bulk).
COMPARE Portugal Laurel (left); Loquat (p.156); Southern Evergreen Magnolia (p.142).
VARIANTS 'Camelliifolia' has leaves curling curiously in hoops, as if treated with weedkiller. Rare.

Several dwarf, pretty forms, such as 'Otto Luyken' (1940), are now very popular ground-cover plants.

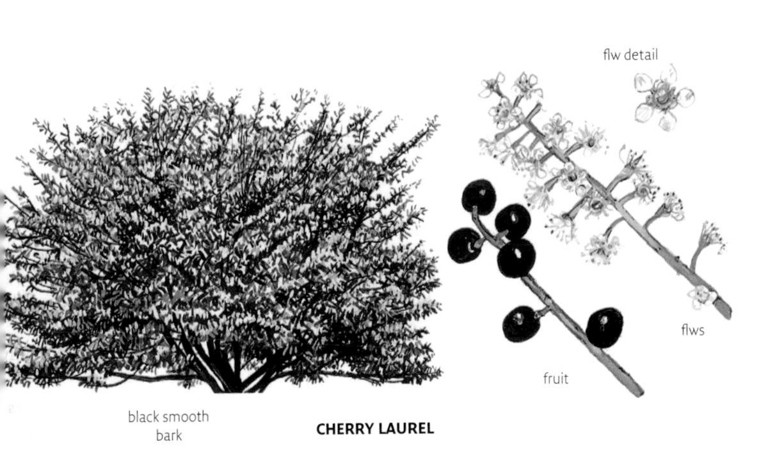

PEA FAMILY TREES

Leguminosae is an enormous family (1 200 species in the *Acacia* genus alone) that is mostly herbaceous. Their leaves are usually compound and seldom toothed; buds are often hidden in the bases of leaf-stalks; and their fruits come in bean-like pods. Nitrogen-fixing bacteria in root-nodules allow many species to colonize poor, sandy soils

Mimosa
Acacia dealbata

(*A. decurrens* var. *dealbata*; Silver Wattle) SE Australia and Tasmania. 1820. The hardiest *Acacia*. Locally frequent in the mildest areas. **APPEARANCE Shape** Often conic with a tilted, wispy tip and light limbs, to 23 m; sometimes broad and leaning with sinuous branches. Very silver trees (var. *alpina*) are hardiest. **Bark** At first glaucous-green; then smooth coppery-brown and minutely wrinkled. Older trees blackish, with fluted trunks. **Leaves** Feathery, doubly compound, to 15 cm; evergreen; individual leaflets only 3.5 × 0.5 mm. **Flowers** Creamy buds stand above the foliage from autumn and open from late winter to spring as dazzling 10 cm branching strings of tiny yellow balls of stamens. Seed-pods to 10 cm; blue-white.

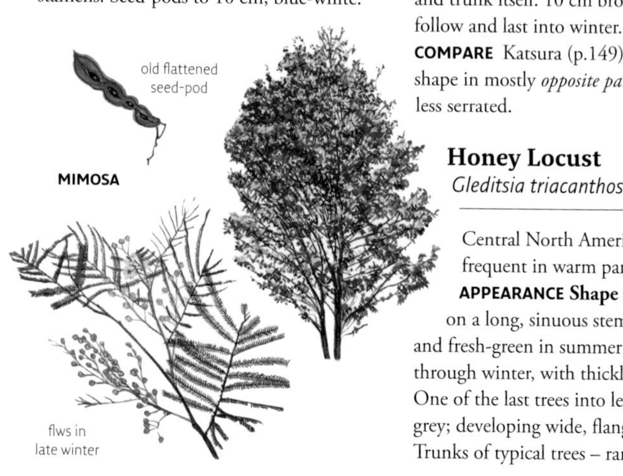

Judas Tree
Cercis siliquastrum

E Mediterranean; long grown in the UK and quite frequent in parks and gardens in warmer parts.
APPEARANCE Shape Low (to 14 m); twisted branches and dense grey-tinged foliage; sprawling and layering given the chance. **Bark** Dark grey; smooth then with *close vertical corrugations*; small, rugged ridges in age. **Shoots** Red-brown. **Buds** Sharp, red; hidden in the bases of the leaf-stalks then rimmed by their grey scars. **Leaves** Kidney-shaped, to 10 cm wide, *untoothed and grey-green*, hairless. **Flowers** Like Sweet Peas, magenta-pink, in late spring with the young leaves; budding endearingly from the bark of the branches and trunk itself. 10 cm brown pea-pods follow and last into winter.
COMPARE Katsura (p.149): leaves of similar shape in mostly *opposite pairs* and more or less serrated.

Honey Locust
Gleditsia triacanthos

Central North America. 1700. Locally frequent in warm parts.
APPEARANCE Shape Slender and open on a long, sinuous stem, to 27 m; dainty and fresh-green in summer but very gaunt through winter, with thickly curling twigs. One of the last trees into leaf. **Bark** Purplish grey; developing wide, flanged ridges. Trunks of typical trees – rare here – carry

JUDAS TREE

umps of vicious 20 cm spines. Thornless rms (f. *inermis*), often selected for straight inks (eg. 'Shademaster'), have naturally en preferred in streets and parks. **Shoots**

HONEY LOCUST

With 3 spines by each 1 mm orange bud in the thorny form. **Leaves** To 20 cm, *singly or doubly compound on the same tree*; the small (2–4 cm) glossy leaflets have *wavering margins* (rarely toothed). The doubly compound leaves have up to 4 pinnae (each like the illustrated leaves but usually smaller) on each side of their green central stalks. **Flowers** In thin, inconspicuous tails, followed (on non-sterile clones) by huge black-brown pods, like squashed bananas, which clatter in the wind. The 'honey' is their pulp, from which beer was once brewed in America.

COMPARE False Acacia (p.198).

Common Laburnum ☠

Laburnum anagyroides

(Golden Chain Tree) Central and S Europe. Long grown in the UK: abundant as an older plant and rarely naturalizing. The hard, dark, greenish timber is valued in cabinet-making, but the whole tree is highly poisonous.

APPEARANCE Shape Irregular, or rather weeping; to 10 m. **Bark** Greenish/coppery brown; smooth for many years, then distantly fissured or slightly corky. **Shoots** Grey-green

COMMON LABURNUM

with *silky hairs.* **Buds** *Silvery-hairy.* **Leaves** *Sparse*; the three 6 cm untoothed leaflets emerge silvery and remain *silky-hairy beneath.* **Flowers** In strings to 25 cm long in late spring; abundant, twisted seed-pods follow.

OTHER TREES Scots Laburnum, *L. alpinum* (damper places and higher mountains in S Europe; 1596), is occasional (least scarce in Scotland); a taller, upright, rather funnel-shaped tree, to 12 m. Its browner bark becomes *ruggedly plated*; shoots are *soon hairless*; leaves rather larger and darker, a deeper green beneath and *almost hairless.* Flowers a fortnight later: longer strings (to 40 cm), though less packed with blossom; the seed-pod's upper seam has *1 mm flanges.*

Voss's Laburnum, *L. × watereri* 'Vossii', is the form of the hybrid of these two tree-species which is *now the commonly planted laburnum*: flower-strings as long as Scots Laburnum's and as crowded as Common Laburnum's. Bark and habit are intermediate; young shoots are soon hairless, the dense leaves are very slightly hairy beneath; the toxic pods *scarcely develop* (a major recommendation wherever they may tempt children).

False Acacia
Robinia pseudoacacia

(Black Locust) E USA. By 1630. Abundan in warm areas, naturalizing by suckers and more rarely, seedlings.

APPEARANCE Shape Gaunt, to 28 m: twisting, much-shattered branches on often straight, slanting stems. **Bark** Grey-brown; soon *very craggy* with long, *de fissures.* **Shoots** Dark red, ribbed: stronger ones with *two spines* by each tiny, scale-less bud (hidden in summer by the base of the leaf-stalk, and then rimmed by its scar). **Leaves** With 9–23 *oval*, untoothed leaflets (each 4 × 2 cm), fresh/bluish green, with a little, soft bristle at the *round*, minutely notched tip; soon hairless. **Flowers** White, scented: laburnum-like cascades at the star of summer after hot years. Dark brown, 10 cm seed-pods hang on in bunches.

VARIANTS Golden Robinia, 'Frisia' (1935) is now the most abundant and perhaps th *brightest* golden broadleaf in gardens in warm areas: large leaflets, clear yellow all

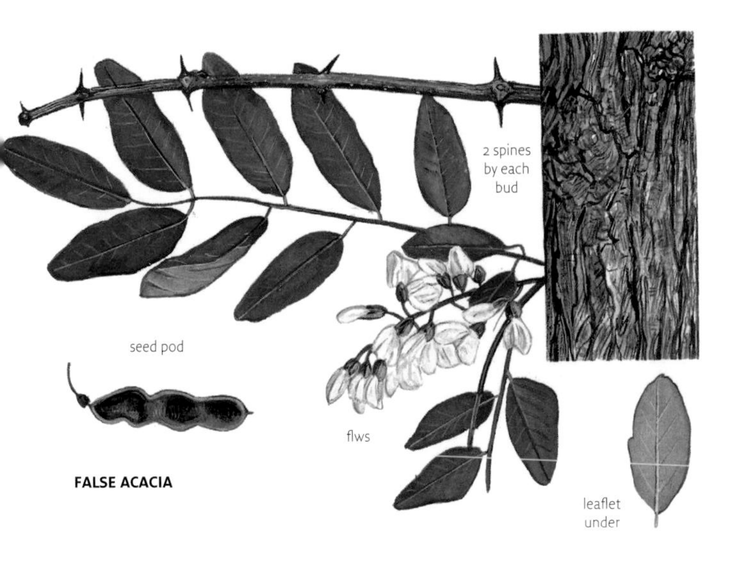

ımmer, darken to gold in autumn. (In very ɔt summers they can fade to pale green.) lender, to 18 m; short-leaved. ('Aurea', by 864 and now very rare, is broader: yellow aves mature lime-green. Type trees flush riefly yellow.)

'Bessoniana' (by 1871) is a locally equent street tree: a good *straight* trunk ypically 5 m) carries a *rounded crown* of eavy, twisting limbs, to 4 m; spineless (except om below a graft) and y-flowering.

Mop-head Robinia, Jmbraculifera' (*c.*1820; ften mis-named nermis'), is frequent: a *izz of often thin, twisting ems* from a graft makes *dense* but fragile low ome, to 9 m; very seldom owering.

'Pyramidalis' Fastigiata'; 1839) is rather re: *Lombardy Poplar-aped*, with few flowers.

Pagoda Tree
Sophora japonica

(Scholar's Tree; *Stryphnolobium japonicum*) China, Korea (long grown in Japan). Quite frequent in warmer areas. 1753 original still thriving – horizontally – at Kew.

APPEARANCE Shape Roughly domed, on heavy twisting limbs: a big, broad tree, to

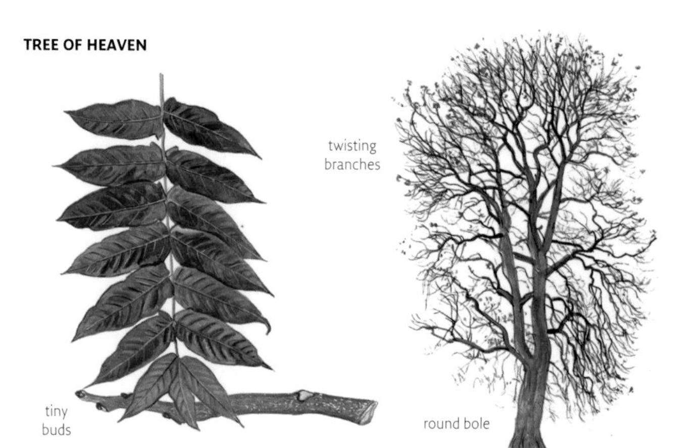

25 m, sometimes recalling Huntingdon Elm (p.133) in winter; rich green in summer with yellow growth-tips. **Bark** Grey-brown; *criss-cross ridges*, less rugged and irregular than False Acacia's (p.198). Shoots *Dark green*. **Buds** 1 mm, hidden in the leaf-stalks then rimmed by their scars. **Leaves** To 25 cm: *9–15* untoothed, *finely pointed* 3–6 cm leaflets. **Flowers** On trees 30 or more years old, showy after hot summers: white (mauve in 'Violacea'), *big branching heads in early autumn*. Seed-pods, to 8 cm, are rare in the UK. Flowers and fruits are much used in Chinese herbal medicine.
COMPARE False Acacia (p.198)

Tree of Heaven
Ailanthus altissima

N China. 1751. (A Moluccan member of the genus is called 'Ailanto', 'a tree reaching the skies'.) Scarce in the wild but aggressively naturalizing in US cities and S Europe. Frequent in warm parts: one of the toughest trees for dry, polluted urban sites. (Family: Simaroubaceae.)
APPEARANCE Shape Tall (to 28 m); stout, twisting branches from a short, column-like trunk. **Bark** Smooth, dark grey, with *fine white vertical 'snakes'*; sometimes slightly corky ridges in old age. **Shoots** Thickly tapering, finely velvety. **Buds** (Above big, pale, heart-shaped leaf-scars) tiny, reddish, domed. **Leaves** Huge, malodorous; flushing late and *red* (cf. Common Walnut, p.95); 11–21 glossy leaflets lightly downy at firs underneath, and around the margin, whi is untoothed except for *1–6 big teeth at th base of each leaflet* (glands on these ooze nectar, attracting ants to combat leaf-eati insects). **Flowers** Usually dioecious. Male in stinking, cream plumes in mid-summe Females, understandably, are the norm, sometimes as grafts: fruits like ash-keys in huge bunches, glamorous and tropical-looking in late summer when – given enough warmth – they ripen apricot, scarlet or ('Erythrocarpum') crimson.

Box
Buxus sempervirens

Europe (including SE England); N Africa; W Asia. Very locally abundant on chalk scarps (place-names suggest a wider forme distribution); naturalized more widely and grown everywhere in garden hedges. The valuable, bone-like timber can sink in wat (Family: Buxaceae.)

PPEARANCE Shape Generally densely ıshy, on steep sinuous stems; to 12 m. **ark** Grey-fawn; close rugged ridges. **eaves** Opposite: small (2–3 cm), convex, ntoothed, often notched at the tip; glossy, athery and evergreen, with an odd sweet nell. **Flowers** Yellow, tufting the shoots om late winter.

OMPARE Phillyrea 'Latifolia' (p.241).

ARIANTS Most are bushy. 'Aurea Pendula' o 8 m; occasional) weeps, with gold- ıargined leaves.

THER TREES Balearic Box, *B. balearica* ılso from SW Spain and Sardinia; 1780), rare: *larger, flatter, matt yellowish leaves* o 5 cm); showier, scented flowers.

tag's-horn Sumach

'hus typhina

North America. *c.*1610. Abundant ı small gardens, and suckering ggressively. (Family: Anacardiaceae.)

PPEARANCE Shape A *miniature, ort-trunked dome* (to 7 m). **Bark** Brown, nely roughened. **Shoots** *Like stags' antlers ı velvet.* **Buds** 4 mm hairy, orange, scale- ss domes; hidden in summer by the base f the leaf-stalk. **Leaves** To 60 cm: up to 25 *serrated*, long-pointed leaflets. They droop then fall individually, often crimson, but may hang into winter. **Flowers** Dioecious. Most garden trees are female: dense, crimson-furred fruiting spires last until spring.

VARIANTS Cut-leaved Sumach, 'Dissecta' ('Laciniata'), has very feathery foliage; occasional (cf. Cut-leaved Walnut, p.95).

HOLLIES

Aquifoliaceae is a widely distributed family dominated by the hollies.

Common Holly
Ilex aquifolium

Europe (including Britain and Ireland); W Asia. Abundant on non-chalky, free-draining soils away from the coldest areas: woods, old commons and hedges, and even exposed moorlands and shingle ridges. Planted everywhere: the sole European member of this genus of 400 diverse species has some claim to be the most ornamental of all.

APPEARANCE Shape Spire-shaped, then irregularly upright, to 23 m; often with densely cascading lower shoots; suckering and sometimes widely layered. **Bark** Brownish grey, finely roughened and frequently with small, round warts. **Leaves** Evergreen, with a high-gloss finish above (pale and matt beneath); massively spined on young plants. The spines deter browsin mammals but are expensive to produce: tall trees have almost entirely unspined, plane leaves and can look puzzling. **Flowe** In late spring. Males have a cross of four yellow-headed stamens; females (on separa trees) have a single style in the middle. **Fruit** Satiny-crimson, toxic berries (whose abundance hinges on how many pollinatir insects were on the wing at flowering time.

COMPARE Highclere Hollies (p.204). Sometimes Phillyrea (p.241): similarly spined but *opposite* leaves. Chinese Tree Privet (p.240): glossy, always unspined leaves. Cork Oak (p.124) and young Holn Oaks (p.123): softer spines and different barks.

female tree

VARIANTS Clones of Common Holly are either male or female, but trees can be sexed with confidence only when in flower as unpollinated females will not berry. Some trees, marketed before they had ever flowered, have names that belie their sex.

Golden Holly, f. *aureomarginata*, is popular in a range of clones: leaves normall shaped, but margined with deep yellow.

Silver Holly, f. *argenteomarginata*, is as common as Golden Holly in various clones its normal-shaped leaves have white/pale creamy-yellow margins.

Hedgehog Holly, 'Ferox', grown since Stuart times and now very occasional. The small leaves have a stubble of spines *across the upper face* towards the tip: a dark, rathe gaunt male tree (to 12 m only) which is onl obvious at close range.

Weeping Holly, 'Pendula' (rather rare), makes a dome of foliage in *dense curtains* (except in dense shade) from a high graft; female.

'HEDGEHOG HOLLY'
COMMON HOLLY
small spines
shiny, spiny lvs
fruit
'SILVER HOLLY'
some lvs have no spines
♀ flws
'WEEPING HOLLY'
♂ flws
some lvs fully yellow
'GOLDEN HOLLY'
some trees are spineless
bark

Perry's Weeping Holly, 'Argenteomarginata Pendula', is one of the best hollies, fruiting freely: the low dome is of brightly silver-edged leaves.

Highclere Holly
Ilex × altaclarensis

The tender Madeira Holly used to be popular in conservatories; in spring the pots were wheeled into the air and Common Hollies sometimes fertilized the females. Their descendants are particularly vigorous, pollution-tolerant trees, with flattened shoots and broader leaves than Common Holly's, but a similar bark and habit. Some cannot be named with confidence. Forms include the following.

Hodgins' Holly, 'Hodginsii' (by 1836), is now occasional: leaves to 10 × 8 cm, thick and glossy, with irregular, small, level spines. Male, its shoots purple in sun. To 22 m; strong single trunk or trunks much swollen where each branch leaves.

'Camelliifolia' has longer, slenderer, finely pointed, high-gloss leaves (to 13 × 6 cm); female, with violet-based petals. A narrow tree (to 20 m), its foliage dense and slightly weeping; the leaves vaguely suggest Japanese Evergreen Oak's, and even Tarajo's.

'Golden King' (female, of course; *c.*1870) is a sport of 'Hendersonii', to which unhappily it readily reverts. Quite frequent, and often densely bushy; its leaves have a few small spines and are matt, but *brightly gold-margined*; 'Lawsoniana' (*c.*1865; occasional also female) instead has a *central marbling of yellow and pale green.*

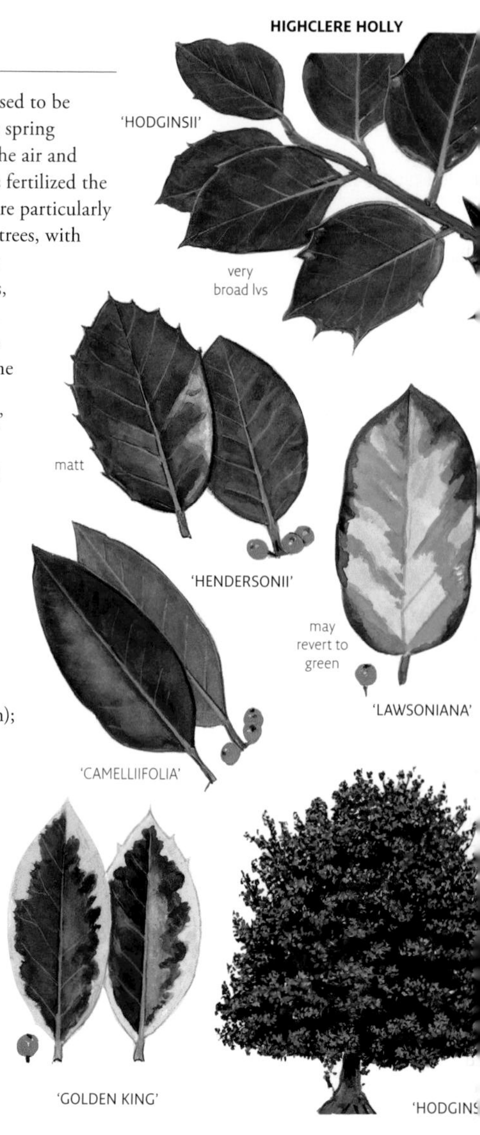

MAPLES

laples (150 species) have leaves in opposite pairs that are very varied in shape, but
ıost typically have five veins radiating neatly at 45° from the base to five lobes.
owers occasionally have showy petals; fruits are bunches/strings of paired 'keys'.
amily: Sapindaceae.)

HINGS TO LOOK FOR: MAPLES

Shape: What shape is the tree?
Bark: What is it like?
Shoots: What colour are they?
Buds: What colour are they? Are they stalked?
Leaves: What shape? Are they hairy or glaucous beneath? Are their teeth whisker-tipped?
Flowers: What colour are they? Are the flowers/fruits bunched (erect/ drooping?) or in tails?

young tree

EY SPECIES 

ield Maple (below): 5-lobed leaves with w, rounded teeth. **Norway Maple** (p.208): -lobed leaves with few, long-pointed teeth. **ycamore** (p.206): 5-lobed leaves with ıany coarse jagged teeth. **Smooth Japanese laple** (p.212): 5- or 7-lobed leaves with any fine teeth. **Cappadocian Maple** .209): 5- or 7-lobed leaves, more or less ntoothed. **Silver Maple** (p.210): leaves ry glaucous beneath. **Paper-bark Maple** .214): trifoliate leaves; red bark. **Box lder** (p.214): 3–7 leaflets; grey bark.

ield Maple 

cer campestre

nglish Maple) Europe, including England ıd Wales; SW Asia; N Africa. Abundant ı rich/heavy soils; frequently planted erywhere.

APPEARANCE Shape Densely twiggy (often almost solid in winter); usually domed when freestanding; to 25 m. **Bark** Pale rather bright brown, with close ridges from the first: slightly corky (cracked into small squares in age) and feeling warm in cold weather. **Shoots** Thin; pale brown, with small grey-hairy buds. They are wrinkly by their second year then may develop corky wings (cf. English Elm, p.130; Sweet Gum, p.151). **Leaves** *Small* (to 10 cm wide); very dark and slightly shiny. The 5 (3) neat lobes have just a few big *rounded* teeth. Rich yellow autumn colour (rarely red). The sap, like that of Norway Maple, Oregon Maple, Miyabe's Maple and Cappadocian Maple and its allies, is *milky* not clear (snap a leaf-stalk and squeeze). **Flowers** In little erect yellow-green posies as the leaves emerge; the green/crimson keys have *horizontal* wings. Some trees are consistently male or female.

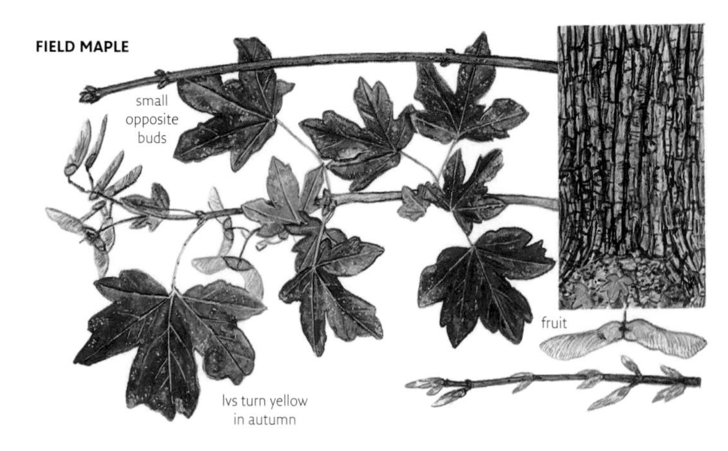

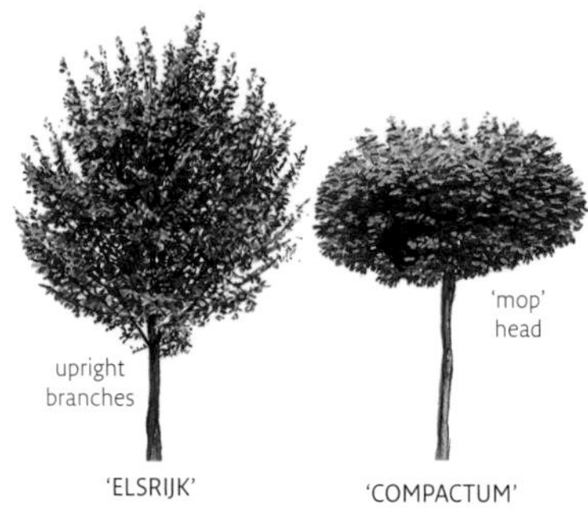

VARIANTS Fastigiate Field Maple, f. *fastigiatum*, is planted occasionally in streets and parks in a number of grafted clones such as 'Elsrijk', though wild trees (especially when crowded) can have closely ascending limbs. No clone is strikingly narrow for long, and all broaden with age.

'Compactum' has a small, globular crown of tiny leaves; rare.

'Pulverulentum' has leaves densely white-stippled. A very bright, ghostly tree, slow and spiky, but rare and tending to revert.

OTHER TREES Montpelier Maple, *A. monspessulanum*, from the E Mediterranean (1739), is in some town parks. Habit very dense; bark *dark grey*; leaves *simply three-lobed*; key-wings *converge*.

Sycamore ✪ ⊕
Acer pseudoplatanus

(Great Maple and – in Scotland – 'Plane') Europe N to Paris; long grown in the UK and fully naturalized: seedlings quickly dominate on rich/heavy soils. Much hated and much hacked: the heavy leaf-fall smothers ground flora, though it does support a high insect biomass – so plenty of birds. Strangely tough (considering its origins): highly valued for its shelter in uplands and on coasts.

APPEARANCE Shape A huge, heavy dome *to 38 m* where it does best (e.g. S Scotland, Kent); twisting, short twigs on straight limbs. **Bark** Pinkish grey: smooth, then by 80 years *shaggy* with small pale grey plates. **Shoots** Greenish grey-pink, stout. **Buds** Big, *green*. **Leaves** Large – to 18 × 26 cm on young trees – with many *coarse, round-tipped teeth*; dull dark green; some tawny down beneath. They flush orange-brown and, in lowlands, drop early, much blackened by Tar-spot Fungus (*Rhytisma acerinum*). **Flowers** Yellow-green, on 6–12 cm *tails*; the keys (green; red in f. *erythrocarpum*) hang from the tail's central stem.

ARIANTS Purple Sycamore, f. *purpureum*, s frequent in gardens and the wild: bark ink-tinged; leaves *green above but mauve/ oyal purple beneath* – a novel maroon- rown as the sun shines through them. The arkest ('Atropurpureum') are grafts.

'Brilliantissimum' is quite frequent: a ow, dense dome (ultimately 15 m); small, harp-lobed leaves unfold *pink-white, fading yellow-green* by summer. 'Prinz Handjery' 1883) is more local: its darker leaves stay *nauve* beneath (cf. 'Nizetii', p.208).

Variegated Sycamore, f. *variegatum*, has eaves *radially splashed* yellow-white; uite frequent and brightly yellowish a distance; bark usually pale grey.

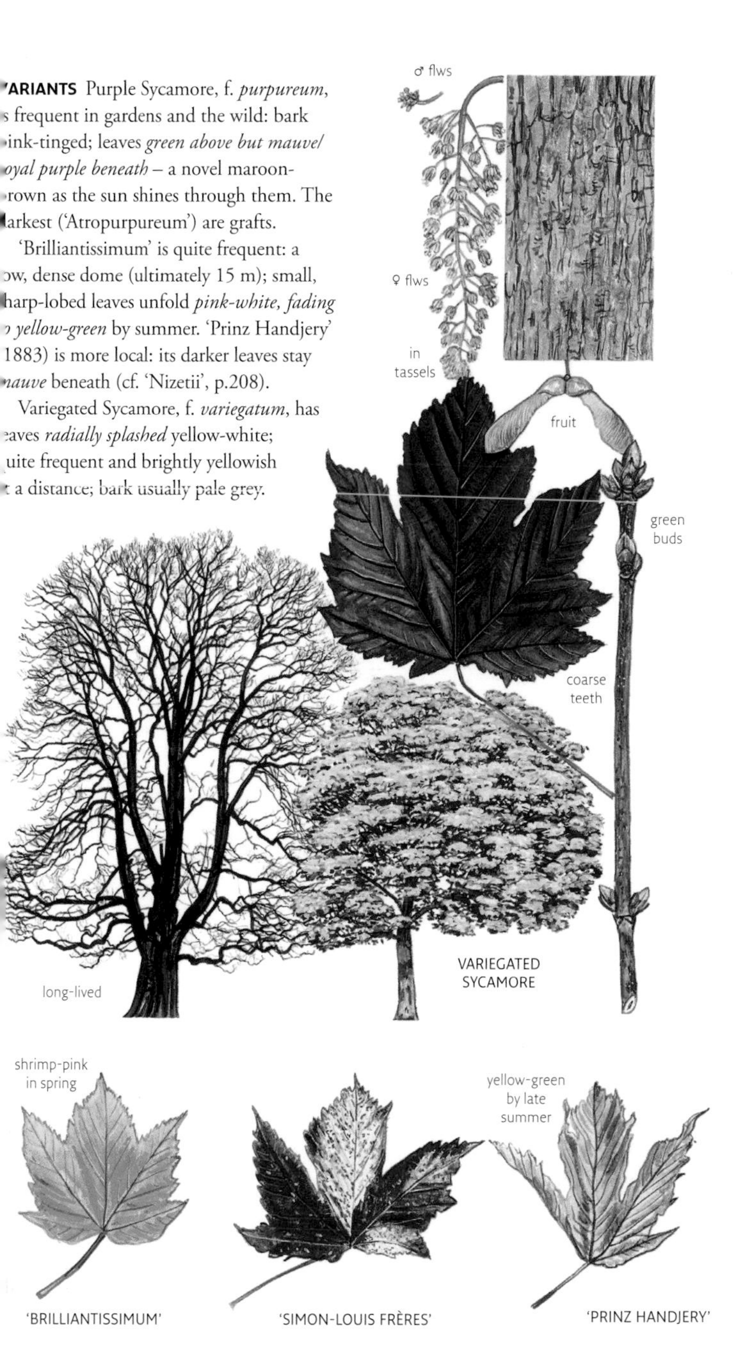

'Nizetii' (occasional; rather narrow) has leaves also *mauve beneath*. Clones include 'Leopoldii' (*c.*1860; yellowish pink/purple splashes in spring), and 'Simon-Louis Frères' (darker leaves flush pink and stay pinkish creamy-green beneath).

Golden Sycamore, 'Worleei', has small leaves with triangular lobes, *bright yellow* then fading slowly to lime-green except at the growth-tips; occasional. Corstorphine Plane,'Corstorphinense' (1600), has leaves of normal size and shape, fading *from gold to deep green in six weeks*; now rare.

Norway Maple

Acer platanoides

Europe (but not Britain/Ireland); the Caucasus. 1683. Abundant; sometimes well naturalized.

APPEARANCE **Shape** A neat leafy dome; to 30 m. **Bark** Pale grey; *closely corrugated* with small regular ridges; oldest trees more rugged. **Shoots** Shiny brown; broad, red-brown buds. **Leaves** Elegant, plane; *hairless* except for tufts under the vein-joints; the *few long teeth mostly* (until tattered by weather) *have whisker-tips*. Yellow autumn colours (rarely purplish then red). Squeezed stalks ooze *milky* sap. **Flowers** In acid-yellow *branching erect posies,* with the leaves.

VARIANTS 'Drummondii' (1903) is one of the brightest and commonest silver-variegated trees (yellow as its leaves open); rather globular and slow, but speeding up a reversions take over.

'Goldsworth Purple' (1936) has *heavy purple* foliage (greener beneath and greenish black by autumn); abundant, to 17 m so

ar. Winter buds saturated purple; *purple* bracts and stalks make the flowers look orange-brown. 'Crimson King' (1946) and 'Faassen's Black' (1936) are hard to separate; 'Crimson Column' is newer and erect.

'Schwedleri' (1870), frequent as a *large* older tree, to 28 m, starts off purple-*red* (with *red* flower bracts/stalks); its leaves turn mauvish *grey-green* by high summer, but show purple autumn colour. 'Reitenbachii' (rare?) is *lower*, and *often burry*: slenderer leaves, more *reddish* green in summer. 'Columnare' is occasional as a younger street tree: leaves short-lobed; steep branches make a narrow, *dense, often irregular funnel/column*, to 25 m. 'Olmsted' (1952) and 'Almira' are broader but shapelier (rare?).

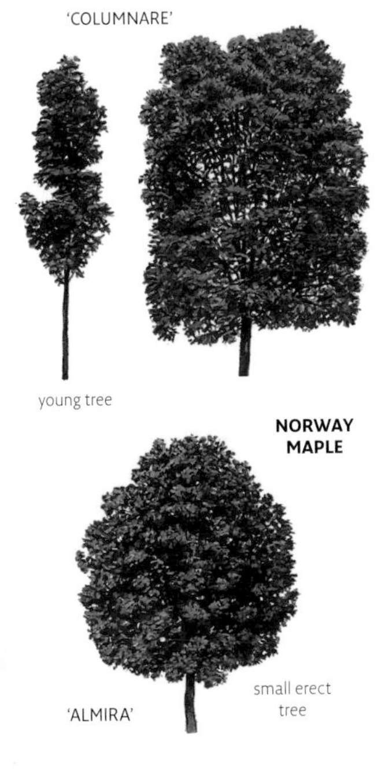

Cappadocian Maple
Acer cappadocicum

(Coliseum Maple; *A. laetum*) Caucasus region, extending E to China (as ssp. *sinicum*). 1838. A frequent street and park tree, closest in appearance to Norway Maple (opposite) but the commonest species whose leaves have consistently *untoothed* lobes; rarely naturalizing by its *suckers*.

APPEARANCE Shape Densely broad-domed on a usually short, often knobbly bole; to 25 m. **Bark** Pale grey: *smooth* then very shallowly fissured. **Shoots** Crimson on strong growths/suckers, bloomed grey; then

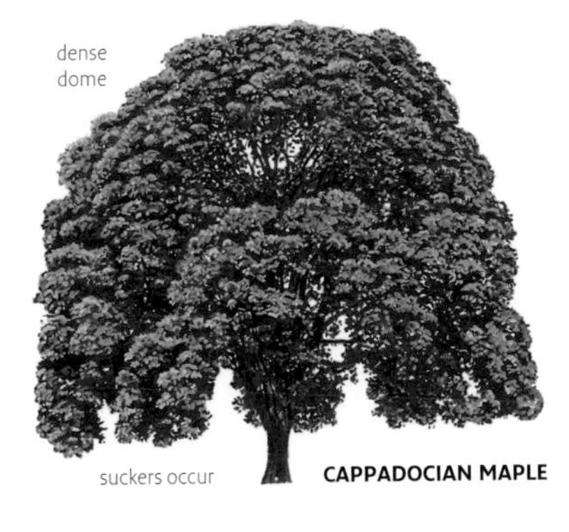

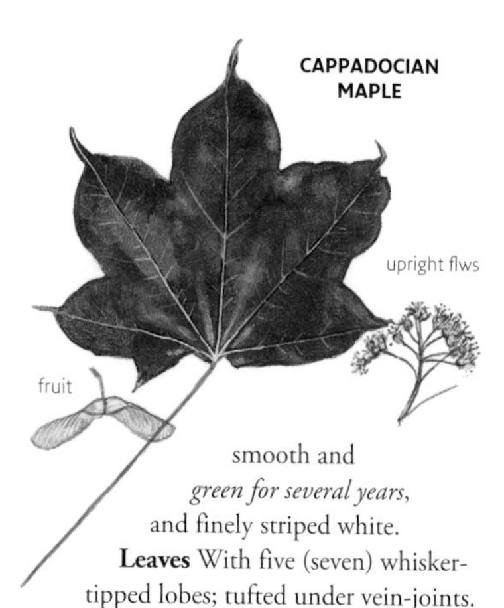

smooth and *green for several years,* and finely striped white. **Leaves** With five (seven) whisker-tipped lobes; tufted under vein-joints. They expand briefly crimson, and are invariably butter-yellow in autumn. **Flowers** Yellow: in erect, branching heads among the leaves in late spring.

VARIANTS 'Aureum' has brilliant whitish yellow younger foliage; this fades slowly to dark olive-green but is still sprinkled with the yellow stars of late leaves when the crowns of type trees become sprinkled with early-colouring foliage. Occasional, to 22 m.

'Rubrum' is a confused name, sometimes applied to the common clone in cultivation but more often to rare grafts with consistently crimson young shoots and leaves retaining pinkish margins. The Chinese ssp. *sinicum*, in collections, has smaller, more deeply and narrowly lobed leaves, a slightly rougher bark, and bright red keys.

ACER MONO

often shaggy bark

OTHER TREES *A. mono* (*A. pictum*; E Asia, 1881) is in many collections, to 18 m. Differs in often *shaggy* whitish bark and twigs *grey-brown and fissured by their second year*; the leaves are downy beneath in the rare f. *ambiguum*.

Shandong Maple, *A. truncatum* NE China, 1881) is also confined to collections, and differs from *A. mono* in its delicate, fan-shaped small leaves which are usually *cut flat across at the base*, and never show more than tufts of hair underneath; they tend to flap in the breeze on their very long stalks.

Silver Maple ✪ ✿
Acer saccharinum

(*A. dasycarpum*) Quebec to Florida. 1725. A very frequent municipal tree.

APPEARANCE Shape An irregular, *airy, willowy* dome: steep limbs and often weeping outer branches from a generally leaning trunk; to 30 m. Vigorous but fragile and short-lived. **Bark** Grey, smooth; then, by 60 years, with shallow but *shaggy* cream-grey plates; often sprouty. **Shoots** Green then red-brown, with clear red/green 8 mm buds. **Leaves** *Five-lobed*, jaggedly toothed; *bright silvery-grey underneath* with some fine down; flushing orange-brown; pale yellow and pink-red in autumn. **Flowers** Wreathed along the shoots before the leaves, *dull* ochre or reddish

VARIANTS Cut-leaved Silver Maple, f. *laciniatum* (including the 1873 clone 'Wieri'),

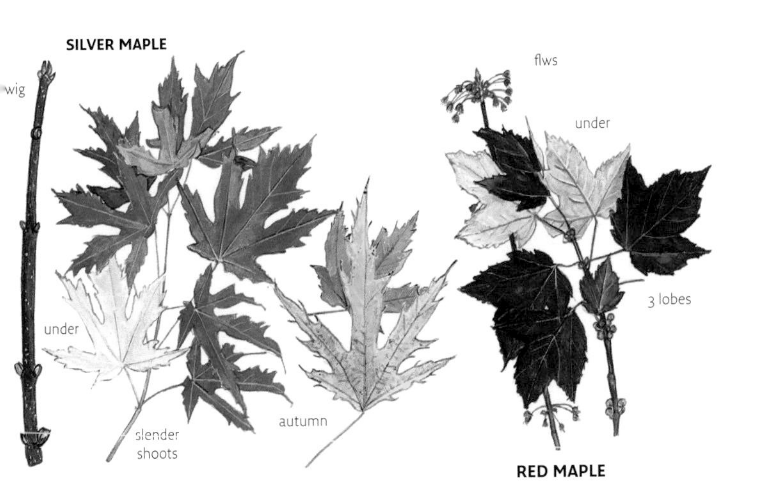

s now the most planted form: leaves *deeply, aggedly cut*; crown often fountain-like.
OTHER TREES Red Maple, *A. rubrum*, from he same habitats, is equally frequent and liffers in 3-lobed leaves.

SNAKE-BARK MAPLES

Snake-bark maples' have indistinctly *stalked buds*; the *leaf-green bark, with vertical streaks,* nay be lost with age. (Spindle, p.246, has lark green bark with fawn 'snakes' as a ounger bush.)

Red Snake-bark Maple

Acer capillipes

(Kyushu Maple) S Japan. 1894. The most frequent snake-bark as a younger planting.
APPEARANCE **Shape** Airily funnel-shaped, on light limbs arching out; to 16 m. The eaves seldom hang. **Bark** Bright; lasting well. **Buds** Purple-red. **Leaves** *Long, with little regular side-lobes* and *sunken parallel veins*; soon almost hairless; 1 mm 'pegs' persist under vein-joints. Rich mixed autumn colours. **Flowers** Yellow-green.
OTHER TREES Père David's Maple, *A. davidii* (China, 1879), is equally frequent; glossy leaves usually *unlobed*.
Grey Snake-bark Maple, *A. rufinerve* (Japan, 1879; rather scarcer) has *white-bloomed* shoots and buds; leaves *matt*, as *broad as long*; rusty hairs under the base are *slowly* shed (no 1 mm 'pegs' in the vein-joints).

Moosewood
Acer pensylvanicum

(Linnaeus' mis-spelling has to be retained.) Nova Scotia to Georgia. 1755. The only American snake-bark maple; it has typical bark. Very occasional.

APPEARANCE Shape Light branches, rising and arching; to 17 m. **Shoots** Green. **Buds** Red/brown. **Leaves** Often *large* (to 22 cm), matt; with *big, usually forward-pointing side-lobes*; rusty hairs underneath are *slowly* shed (*no* 1 mm 'pegs' in the vein-joints); yellow autumn colours. **Fruit** Abundant keys, the nuts *flattened*.

COMPARE Grey Snake-bark Maple (p.211): often smaller, spreading side-lobes; rounded nuts.

VARIANTS 'Erythrocladum' has *pale, clear crimson* winter shoots, and gold younger bark with *red and white 'snakes'*; unfortunately a sickly tree, and now very rare.

Smooth Japanese Maple
Acer palmatum

('Acer') China, Korea and Japan, where long cultivated in many generally bushy cultivars. 1820. A singularly elegant plant, quite frequent in mostly larger gardens; seedlings spring up freely.

APPEARANCE Shape Rounded, to 15 m: light limbs from a short bole and dainty horizontal sprays of dense, star-like, moss-green leaves. **Bark** Smooth and brown-grey: faint white snakes at first, then some muscular ripples. **Shoots** Slender, bright red/green, *all tipped by two* crimson/green buds. **Leaves** 4–7 cm (but to 12 cm in ssp. *amoenum* and

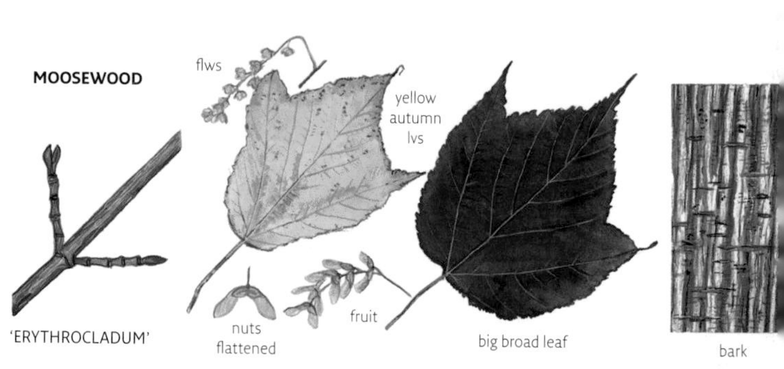

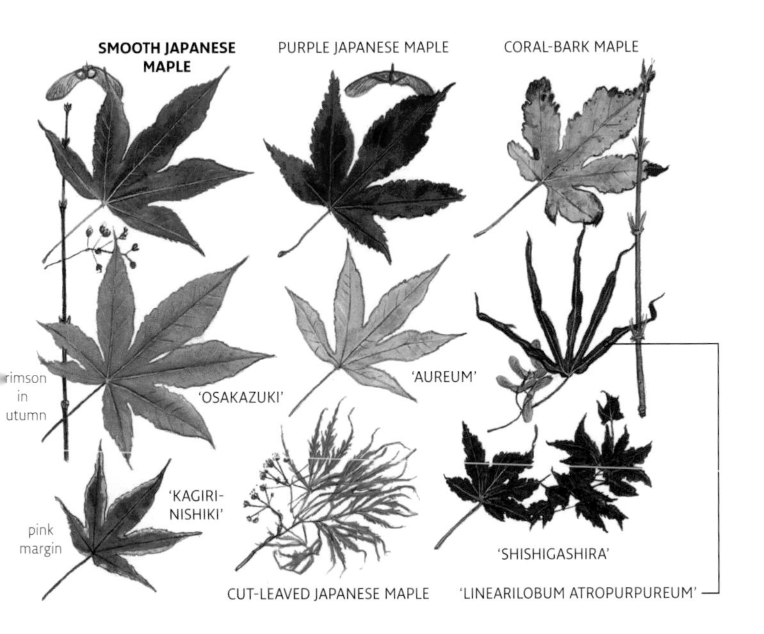

he widely grown var. *heptalobum*); hairless xcept for tufts under the vein-joints; the five r seven (rarely nine) deep lobes are *finely and harply* but seldom doubly serrated. **Flowers** Deep red, in small spreading heads, turning rees amber as the green leaves flush. **Fruits** ale red; erect/drooping clusters.

ARIANTS 'Osakazuki' (a Japanese selection f var. *heptalobum*; 1861) has larger leaves seldom five-lobed), flushing deep pink so hat the crown is maroon-tinged through ummer, with reliably *blood-red autumn olours* (type trees usually turn pale orange); ruits *rich* red. Quite frequent.

Purple Japanese Maple, f. *atropurpureum*, s quite frequent and crops up among eedlings. Leaves bronze, purplish, or a uperb deep crimson in selections like Bloodgood'; paling to scarlet in autumn.

Coral-bark Maple, 'Sango-kaku', has stiffly sing *brilliant red* winter shoots; the small, ightly yellow leaves turn gold in autumn.

Cut-leaved Japanese Maple, f. *dissectum*, as finely feathery foliage and usually grows as a miniature mushroom; to tree height in shade. Very frequent in many selections (cf. cut-leaved Norway Maples, p.208). 'Dissectum Atropurpureum' has soft, silvery-purple foliage.

'Hagoromo' ('Sessilifolium'; rare) was once considered a different species: its leaves (*very shortly stalked*) often have *three or five jaggedly lobed leaflets*. A slender, rather upright tree, to 14 m.

'Linearilobum' is rare: deep, narrow *finger-like* lobes (not themselves dissected); 'Linearilobum Atropurpureum' ('Atrolineare') is the purple version.

'Shishigashira' ('Ribesifolium'), *narrow, stiff and very dense*, has *much-twisted and cut dark leaves*.

'Aureum' has soft yellow leaves (gold in autumn); very rare.

'Albomarginatum' has delicately white-margined leaves; rare; 'Kagiri-Nishiki' ('Roseomarginatum') has five-lobed leaves with fine pink margins (fading creamy-white); tall-growing but readily reverting.

PAPER-BARK MAPLE

Paper-bark Maple
Acer griseum

Central China. 1901. Rather occasional, but in many gardens and some parks for its unique bark. The seeds often mature without fertilization and are infertile, and the difficulty of raising seedlings helps to explain why this gorgeous tree is still not commoner.
APPEARANCE Shape A low, neat, finely twiggy dome (gaunt when unhappy); to 15 m. **Bark** *Papery* scrolls from the first: *rich cinnamon-red* and chocolate-brown, sometimes blue-bloomed. Odd trees can develop harder curling scales, a dull orange-brown. **Leaves** With *three leaflets*, each roundly toothed/lobed; dark grey-green above and hairy; dense blue-white down beneath, and on the pink/red stalks. They flush very late, pale orange; usually orange and crimson autumn colours. **Flowers** Usually in 3s.

Box Elder
Acer negundo

(Ash-leaved Maple; *Negundo aceroides*) Across North America; sometimes tapped for maple syrup. By 1688. Very frequent.
APPEARANCE Shape Broad, *untidy*: leaning stems and steep younger limbs. To 18 m. **Bark** Pale grey; burred, sprouty; *irregular networks of ridges*. **Shoots** Rich *plastic-green*, variously bloomed. **Buds** Silky-white. **Leaves** With *three to five* (rarely seven or nine) *leaflets*, each deeply/haphazardly lobed; velvety on downy shoots in var. *californicum* (W coast); little autumn colou **Flowers** Dioecious: in showy hanging plumes before the leaves.
COMPARE Smooth Japanese Maple 'Hagoromo' (p.213). Larger and much coars than the (consistently) trifoliate maples.
VARIANTS var. *violaceum* covers mid-Western trees with *strongly purple-bloomed* but hairless shoots. Foliage *darker* and heavier; males with spectacular salmon-pin flowers. Occasional.

'Variegatum' (1845; female) has brightly white-variegated leaves and white-winged fruit. Much planted in the mid-twentieth century but quickly reverting: today seen generally as a graft with *white-bloomed shoots*, yellowish, *slightly blotchy-green* foliag and the odd milk-white sprout.

'Elegans' has glossy, boldly yellow-margined leaves and blue-bloomed shoots; probably also reverting quickly as no big trees are now known.

'Flamingo' has white-variegated leaves unfolding pink; rather rare as a small tree, but less unstable.

'Auratum' (1891) has golden then pale green leaves; rather rare. A *female* clone, prone to revert.

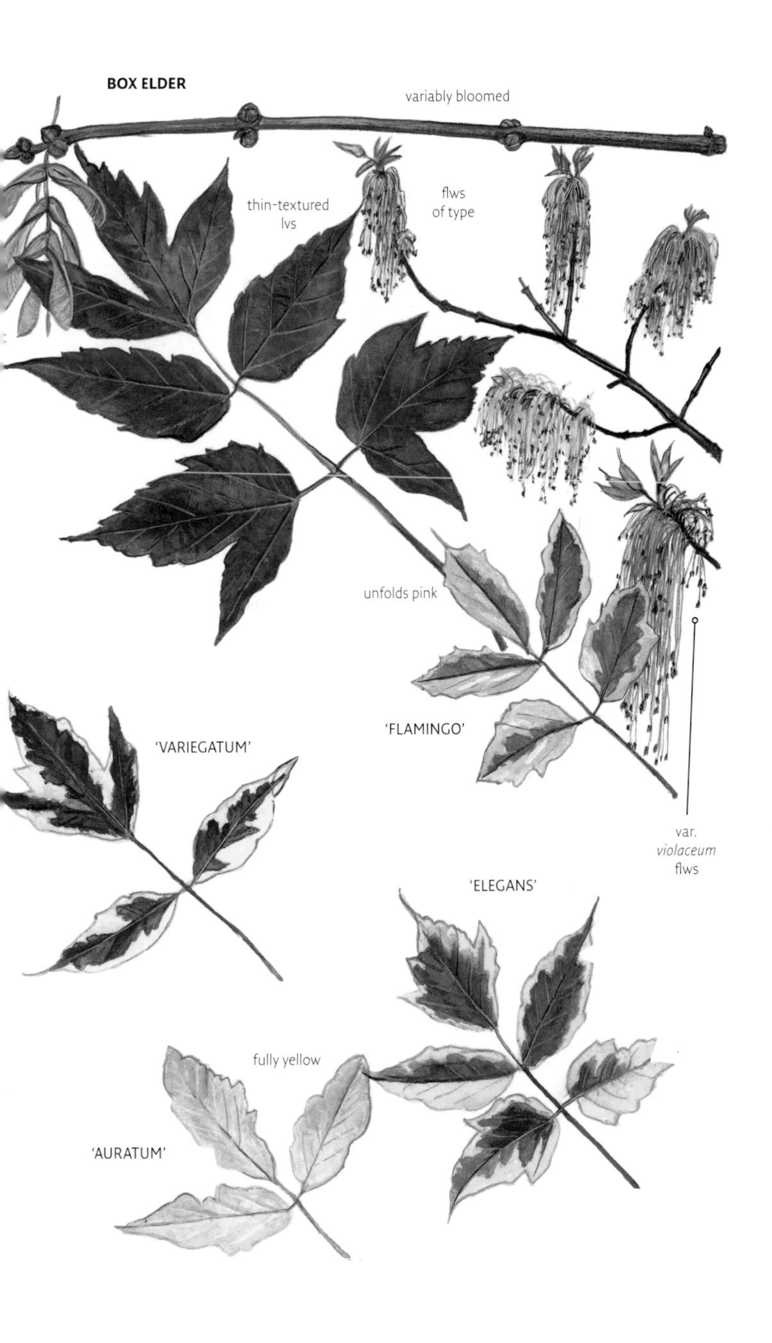
BOX ELDER
variably bloomed
thin-textured
lvs
flws
of type
unfolds pink
'FLAMINGO'
'VARIEGATUM'
var.
violaceum
flws
'ELEGANS'
fully yellow
'AURATUM'

HORSE CHESTNUTS

The 25 horse chestnuts and buckeyes all share compound, hand-like leaves. (Family Sapindaceae.)

Things to Look for: Aesculus

- Bark: Does it remain smooth?
- Leaves: How many leaflets? Are they stalked? Hairy beneath? Doubly toothed?
- Flowers: Colour? Are the petals hairy? Sticky?
- Fruit: Are the conker-husks prickly?

Key Species ✪

Horse Chestnut (below): usually 7 unstalked leaflets. **Indian Horse Chestnut** (opposite): usually 7 stalked leaflets. **Red Horse Chestnut** (p.218): 5 scarcely stalked leaflets. **Yellow Buckeye** (p.219): 5 distinctly stalked leaflets.

Horse Chestnut ✪ 

Aesculus hippocastanum

Growing freely from conkers in the UK and abundantly planted since 1616, but confined in the wild to mountains in N Greece/Albania. Its name probably derives from its role in horse-medicine and its planting by smithies (and from association with the unrelated Sweet Chestnut, whose fruits have a similar design).

APPEARANCE Shape An often narrow dome of short, twisting twigs. The main limbs shatter easily, particularly when rain weighs down the dense foliage. To 39 m; potentially very long-lived. **Bark** In youth, smooth and pink-grey; red-brown and coarsely scaly by 80 years. **Shoots** Thick, red/grey. The pale leaf-scars are horseshoe-like (three 'nail-holes' one side, four the other, where sap ran to each leaflet). **Buds** Deep red-brown, *sticky* (cf. Sargent's Rowan, p.168); end one huge. **Leaves** With *seven (five or six) stalkless* leaflets, jaggedly/double-toothed; unfolding early and by late summer often browned by fungal rusts. **Flowers** The showiest of any tall tree's; the yellow basal blazes can turn red after

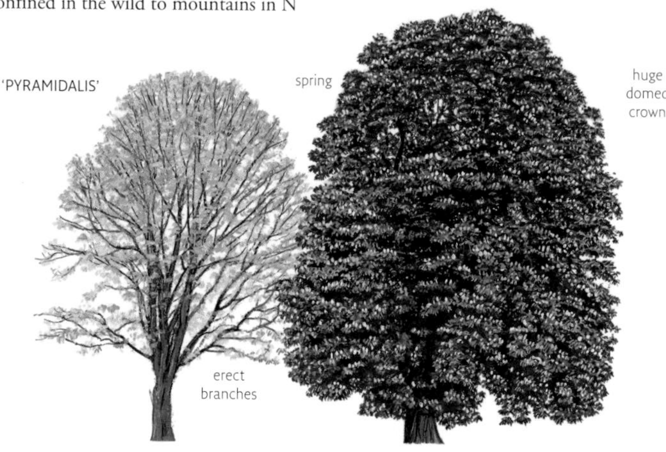

HORSE CHESTNUT

ɔllination. **Fruit** Many conker-husks on each ee bear *sharp spines*.
ARIANTS 'Baumanii' (1820) has whiter, umpier candles of *double* flowers, so no ɔnkers; quite frequent. It grows almost ertical central limbs, *densely clothed* in ather small, cupped leaves.
'yramidalis' is rare: *stiffly upright* branches ıake a neat, narrow tree with a pointed top.
Hampton Court Gold' has evenly yellow-reen foliage (many sickly type trees are lotchily yellow); 'Honiton Gold' is taller: oth are very rare.
)igitata' has very *narrow* leaflets (often only three) on a flattened, leaf-like main stalk; f. *laciniata* has *feathery*, cut leaves; both are feeble and very rare.

HAMPTON)URT GOLD'
'DIGITATA'

Indian Horse Chestnut

Aesculus indica

NW Himalayas. 1851. A splendid tree which is still distinctly occasional; often seeding abundantly.
APPEARANCE Shape A fine dome on straightish limbs, to 26 m; sometimes a giant bush. **Bark** Smooth, pink-grey; distantly scaling with age. **Buds** Green/pinkish red and sticky; weaker shoots end in a pair. **Leaves** With five to nine (usually

INDIAN HORSE CHESTNUT

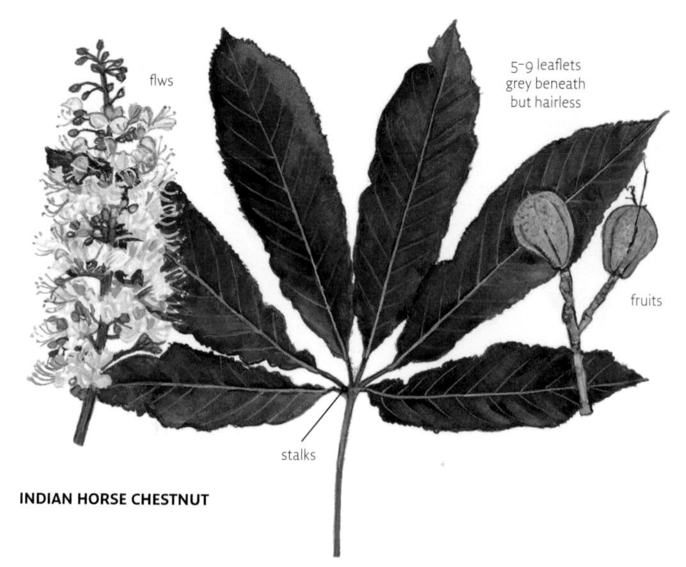

INDIAN HORSE CHESTNUT

seven) *slender*, finger-like leaflets each on *1 cm reddish stalks*; glossy and *very dark* above; pale grey beneath but *hairless*. They open briefly *tomato-red* but – the tree's one limitation – show little autumn colour. **Flowers** Pale pink (with yellow blazes turning crimson), in long, elegant candles, from late spring into *early summer*. 'Sydney Pearce' (1928; rare; grafted trees) has many rich red and even *purple* blazes. **Fruit** Conkers black-brown, in leathery, spineless husks.

Red Horse Chestnut

Aesculus × carnea

A cross of Red Buckeye with Horse Chestnut (by 1818), which has since doubled its chromosomes so breeds true. An abundant plant of rather endearing ugliness.
APPEARANCE **Shape** Low (rarely 20 m); very twisting, often weeping branches. **Bark** Smoother for longer (with prominent lenticels) then less closely/shaggily ridged than Horse Chestnut's; often much cankered. **Buds** Greyish, *scarcely sticky*; weaker shoots end (like the buckeyes') in a pair of buds. **Leaves** Dark, *crumpled* but slightly glossy; *five* (six or seven) leaflets, each *jaggedly* toothed and scarcely stalked; usually smaller than Horse Chestnut's. **Flowers** *Dull crimson* ('*carnea*' means 'mea coloured', which is a bit unfair), in dumpy rather radiating candles. **Fruit** The conker-husks have few or no spines.

RED HORSE CHESTNUT

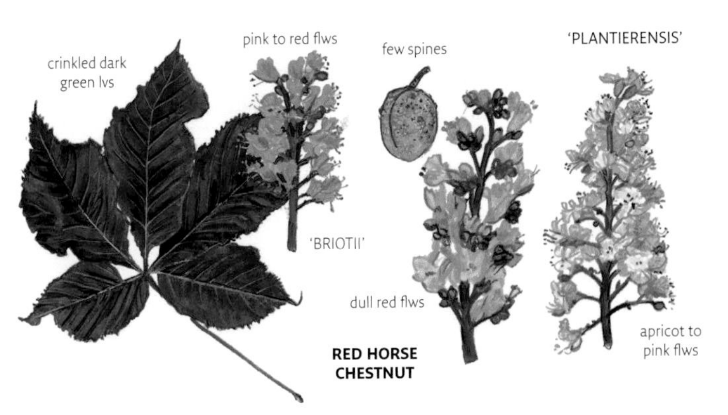

ARIANTS 'Briotii' (1858) is now the ımmonly planted clone: a shapelier dome ɔ 25 m); *glossier,* less-crinkled leaves; ndles variably *bright* red-pink (often one ɔwer with a yellow blaze next to one with a agenta blaze).

'Plantierensis' (*A. × plantierensis*), locally ccasional, is a back-cross with Horse hestnut: a low-domed tree whose long aflets are shaped like Horse Chestnut's but inkled and shiny and usually in 5s, with *ft apricot and salmon-pink flowers* in huge, ose candles – the finest *Aesculus* of all in ossom. It *does not fruit.*

Yellow Buckeye

Aesculus flava

(Sweet Buckeye; *A. octandra*) E USA. 1764. Rather occasional: parks and gardens in warmer areas.

APPEARANCE Shape Narrowly domed, to 26 m, or irregular; often *twisting* branches. **Bark** Pink-grey; smooth (prominently lenticellate), then usually with big, harsh, curved scales. **Buds** Pale pink-brown, non-sticky; weaker shoots end in a pair. **Leaves** With five (three or four) *elegant*, smooth, slightly glossy, fresh-green leaflets each on

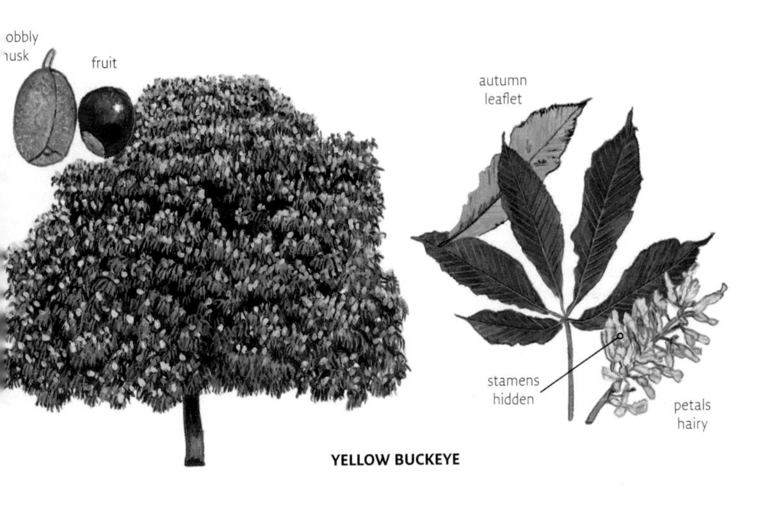

a *1 cm stalk*, and remaining *more or less downy beneath*. A tree grown firstly for its foliage; bright orange-red in autumn. **Flowers** In meagre yellow candles among the fresh leaves – piquant, not spectacular (red in the rare f. *virginica*, which is confined in the wild to W Virginia). The tubular, *long-hairy* petals lack Red Buckeye's sticky glands, and (unlike Ohio Buckeye's) are long enough to hide the stamens. **Fruit** The leathery conker-husks always *lack spines*.

GOLDEN RAIN TREE

Golden Rain Tree
Koelreuteria paniculata

(Pride of India) China, Korea, Japan. 1763. Occasional in warmer parts; sometimes seeding freely. (Family: Sapindaceae.)
APPEARANCE Shape Densely domed, on thick, rising and twisting branches; to 15 m. Matt, very dark foliage. **Bark** Brown; close, ultimately craggy, interlacing ridges. **Shoots** Pale coppery-brown, the short, sharp buds standing above *raised black-rimmed leaf-scars*. **Leaves** Variously subdivided but unmistakable; flushing garish red then amber-pink. **Flowers** Dramatic, 30 cm mustard-yellow plumes festoon the crown in high summer. The equally eye-catching *pinkish bladders*, like Chinese lanterns, each hold 3 pea-sized seeds.

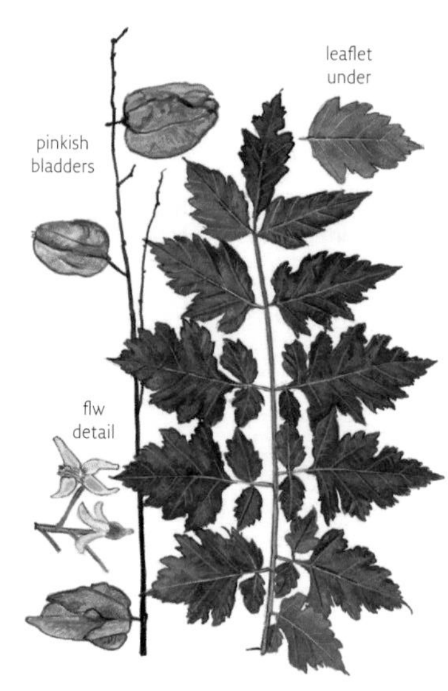

GOLDEN RAIN TREE

Purging Buckthorn 
Rhamnus cathartica

Europe (including England, Wales and Ireland); W and N Asia. Frequent but inconspicuous in scrub and woodland *on calcareous soils*; scarcely planted. This and Alder Buckthorn, its counterpart on acidic soils, are the food-plants of the Brimstone Butterfly, the sulphur-yellow males often being the most conspicuous evidence of the plant's proximity. Its bark and fruits (skin irritants) were once used to make a yellow dye. (Family: Rhamnaceae.)
APPEARANCE Shape Twiggy, often bushy, on a short but sturdy stem; exceptionally to 15 m. Side-shoots often end in Blackthorn like spines. **Bark** Dark brown; soon *shaggily scaly*. **Shoots** Slender, straight, grey-brown. **Buds** Conic, black-brown; *appressed; usually in opposite pairs but sometimes staggered*. **Leaves** Dark, about 6 cm long: with incurving veins like Common Dogwood's (p.233) but *finely round-toothed*. Usually hairless; yellow in autumn. **Flowers** Small yellow-green stars, clustered at the base

new shoots. **Fruit** 6 mm black berries
owerfully purgative and once widely used
a kill-or-cure medicine).
MPARE Spindle (p.246): narrower
ves. Wild Crab (p.171): similar jizz, but
nsistently alternate foliage.

lder Buckthorn
angula alnus

hamnus frangula) Europe (including
gland, Wales and Ireland), E to Siberia;
Africa. Locally frequent in wet heathlands
d swamps, or as a straggling plant in
idic woodlands; scarcely planted. A close
y of Purging Buckthorn, though it looks
ry different. Its timber – a startling lemon-
llow – once provided the finest charcoal
r gunpowder. The sap is intensely bitter
d irritant.

APPEARANCE
Shape Spire-like when young, on elegant, straight, light stems, but scarcely to tree size. **Bark** Smooth, dark grey. **Shoots** Straight, very slender; purple-brown; elongated lenticels make *fine white streaks.* **Buds** Alternate, *scaleless* – 3 mm tufts of orange fluff. **Leaves** Small (about 5 cm), flat; matt fresh green; *untoothed*. They are tapered at the base but *blunt-tipped* (except on strong growths), like those of Common Alder (p.103), with which it often grows. **Flowers** In tiny green clusters. **Fruit** Red berries, ripening purple; toxic.
COMPARE Smoke-bush (*Cotinus coggygria*): similar foliage.

small shrubby tree

LIMES

Malvaceae is a diverse family including one main group of trees – the limes. Limes (30 species) have heart-shaped leaves that are typically bulged more on one side of th slanting stalk than the other; there is no big end-bud. Their flowers, with showy and aerodynamic bracts, are deliciously scented. Leaf aphids often drip sticky 'honeydew' through the summer.

THINGS TO LOOK FOR: LIMES

- Shoots: Are they hairy/felted? How red are they?
- Leaves: Are they glossy? Downy/felted beneath? Are there tufts (what colour?) under the vein-joints? Is the leaf-stalk downy? How long is it?

Broad-leaved Lime
Tilia platyphyllos

Europe (including England and Wales); SW Asia. A rare native from the Pennines, Wye Valley, Cotswolds and SE downland scarps. Planted abundantly for some centuries, though less than Common Lime.

APPEARANCE Shape Tall-domed; to 42 m. **Bark** Greyish, with often clean criss-cross ridges; seldom sprouty. **Shoots** Grey-green (redder in sun; clear red in the rare 'Rubra'/'Corallina', which has dense, bright foliage); *fine hairs* wear off through winter. **Buds** Fat, the *three scales* grey/dull red, with sparse hairs. **Leaves** Often *dull, dark* green and *softly furry*; on *hairy stalks* in N European trees (ssp. *cordifolia*); hairy on under the veins in the central European ssp *platyphyllos*; nearly hairless in the E Europe

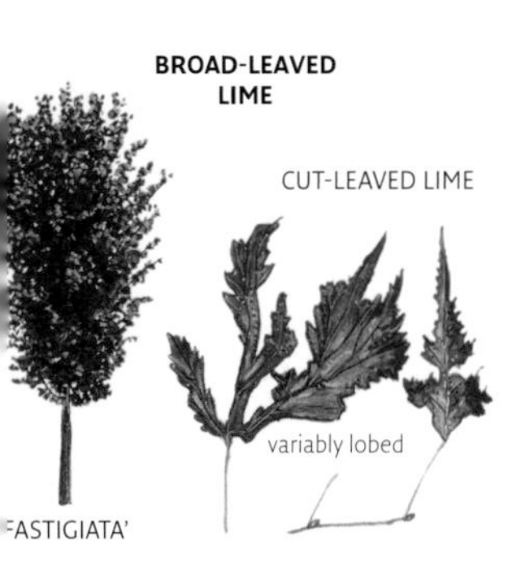

. *pseudorubra*. To 15 × 15 cm; the sides
droop. **Flowers** Hanging, in mid-
nmer; three to six per bract. **Fruit** *Often*
ngly five-ribbed, downy.
MPARE Common Lime (p.224). Silver
ne (p.225): *furry* buds in winter.
RIANTS Cut-leaved Lime, 'Laciniata'
spleniifolia'), has small, very variably
ed leaves; a narrow tree (to 22 m),
vering profusely.
'Fastigiata', very occasional as a young
, has close, steep limbs making a narrow,
nted dome; 'Orebro' (1935; rarer) is more
cefully conic.

Small-leaved Lime

Tilia cordata

(Pry Tree) Europe (including England N to Cumbria, and E Wales); the Caucasus. Locally abundant in old woods and hedges. Pollen deposits show that 5000 years ago this was the dominant tree in the NW European lowland 'Wildwood'; but it is vulnerable to grazing by stock and needs summer heat to ripen its seeds, and is now absent in the wild from many counties. Some giant coppice stools, surviving from prehistory, may be England's oldest 'trees'. Occasional as an older planted specimen, but now in fashion.

APPEARANCE Shape Domed, to 38 m. **Bark** Grey/buff; craggier than Broad-leaved Lime's; often very sprouty. **Shoots** Quickly *hairless* and (in sun) shining red. **Buds** Fat, hairless, with one big and one small scale – like boxing-gloves. **Leaves** Small (8 × 8 cm); usually sturdily *flat-surfaced*; hairless except for *rufous* tufts under vein-joints. Underside matt and slightly silvered (can be glossier green in shade/on saplings). **Flowers** Profuse: *spreading at all angles* in high summer, so that the crown turns creamy yellow; 5–11 per bract. **Fruit** *Hairless; scarcely ribbed.*

COMPARE Common Lime (overleaf): fawn tufts under leaves; drooping flowers.

Common Lime

Tilia × europaea

(*T. × vulgaris*) A hybrid of Broad- and Small-leaved Limes, very rare in the UK in the wild (more widespread on the Continent), but from the early 17th to the mid-twentieth centuries the most planted lime: the liberal sprouts made propagation cheap. In many high streets, submits cheerfully to annual lopping.

APPEARANCE Shape Domed or, in one common clone, *columnar with close, vertical limbs*. The tallest wild broadleaf, to 46 m; the small leaves near the top make it look even bigger. **Bark** Pale grey-brown, irregularly ridged; some clones envelop themselves in sheaths of sprouts, ideal for nesting birds. **Shoots** Red in sun; soon hairless. **Buds** Fat, like boxing gloves: one big and one small reddish scale *fringed* with fine hairs. **Leaves** Typically 10 × 10 cm, *flimsy*, soon more or less hairless except for *buff* tufts under the vein-joints (cf. Small-leaved Lime's more *rufous* tufts). Underside usually shiny pale green, but matt and slightly silvered on the shoots (in sun) of

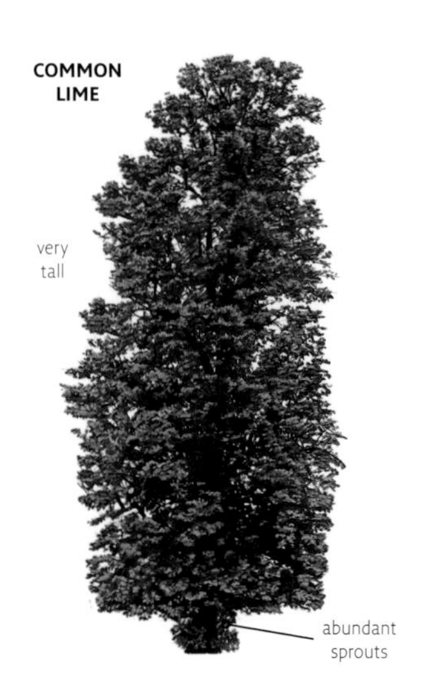

one clone; stalks soon hairless. **Flowers** Hanging, in high summer; four to ten pe bract. Fruits *Downy, but only faintly ribbe*

VARIANTS Golden Lime, 'Wratislaviensis (*c.*1890), is strangely rare: flushing green, then *gold*; later dark but still sprinkled wi yellow late growths.

Crimean Lime

Tilia × euchlora

(Caucasian Hybrid Lime) *T. cordata × dasystyla*? – in the Crimea by 1860. Frequently planted: the *glossy foliage* is nc aphid-friendly.

APPEARANCE Shape The common clone is soon unmistakably ugly: a *straight trunk* degenerates halfway up into a *tangle of curling limbs*, above which fine branches make a *narrow*, dense dome to 20 m. **Ba**

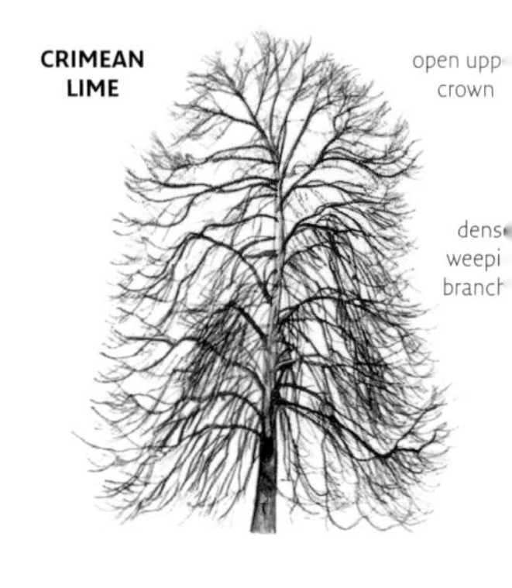

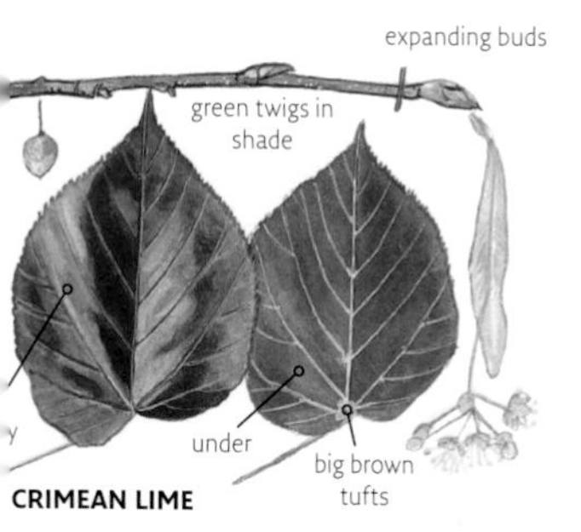

rk grey; smooth; widely fissured with . **Shoots** *Lime-green* (*amber*-red in sun by ater), finely downy. **Buds** Green/reddish. **aves** Plane, elegantly pale-toothed, on rless stalks; *big brown tufts* under the 1-joints. They yellow, one by one, early in umn. **Flowers** Late; rich gold and headily nted. **Fruit** Downy, shallowly 5-ribbed. **MPARE** Common Lime (opposite). ver Pendent Lime (below): slightly similar nching.

lver Lime

a tomentosa

ngary to SW Russia and NW Turkey. 57. Quite frequent in warmer parts, ecially as a younger tree. **PEARANCE** **Shape** Precisely domed,

usually on straight, steeply radiating limbs, to 32 m. The heavy, sombre foliage, in *level plates*, is lit up by the silvery leaf-backs. **Bark** Dark grey-brown; coarse but shallow criss-cross ridges. Old specimens are often grafted. **Shoots, Buds** *Finely but closely grey-downy*. **Leaves** With large teeth (and rarely pointed lobes at the 'shoulders'); *white-woolly* beneath. The minutely silver-scurfy stalk is *less than half as long* as the leaf. **Flowers** Late, seven to ten per bract; very fragrant, with a hint of soap-powder; a sugar in the nectar intoxicates and can kill bumble bees. **Fruit** five-ridged, white-hairy. **VARIANTS** Silver Pendent Lime, 'Petiolaris' (*T.* × *petiolaris*; of obscure origin, by 1842), is slightly more frequent and possibly the most ornamental (and vigorous) lime. Fountain-like crown, to 33 m, with *weeping side-branches from heavy, crooked main limbs*; leaf-stalk (much) *more than half as long* as the leaf. Always grafted (usually on the sprouty Common Lime) and awkwardly outgrowing the stock; good yellow autumn colours.

Kohuhu

Pittosporum tenuifolium

New Zealand. *c.*1850. Quite frequent in mild areas; sometimes seeding. (Family: Pittosporaceae.)

APPEARANCE Shape To 17 m but densely bushy in colder parts, on many slim straight rising stems. **Bark** Dark grey-brown; smooth. **Leaves** Foliage a favourite in floristry: purple-brown shoots carry thin, softly sea-green, evergreen 5 cm leaves, as undulant as crisps. **Flowers** Purple-brown honey-scented, 7 mm bells stand by the le stalks in late spring.

VARIANTS 'Silver Queen' has misty-grey leaves finely margined in silver; frequent. 'Warnham Gold' has soft yellow foliage; rare. 'Purpureum' has green young leaves, maturing a saturated blackish purple; occasional.

Tamarisk

Tamarix gallica

(*T. anglica*) N France to N Africa, near the Atlantic. Long naturalized in the UK (possibly native) and abundant in gardens on coasts; occasional inland. (Family: Tamaricaceae.)

APPEARANCE Shape Broad and bushy, on a gnarled, sprouty stem; to 8 m. **Bark** Brown; stringy with close vertical ridges. **Shoots** Red; willow-like but knobbly with close, sharp buds. **Leaves** In light green sprays 1 mm thick – Cypress-like, but wit spiralling not paired scale-leaves, and shed in winter. **Flowers** Pink (rarely white), spraying *from the young shoots in summer.*

Elaeagnaceae is a family of 50 silver-scaly trees and shrubs. Their roots, like alders and many pea-family trees, carry nodules of nitrogen-fixing bacteria allowing them to grow in poor soils.

Sea Buckthorn

Hippophae rhamnoides

Eurasia (including British coasts from the East Lothian S to Dungeness). Planted and aggressively suckering elsewhere on sand-dunes; a frequent, *intensely silver* garden plant.

APPEARANCE Shape Very spiky and spiny, to 10 m (but columnar in parts of its range); black and stocky in winter. **Bark** Dark grey-brown; shaggy, willow-like edges. **Shoots** *Silver-scaled*, closely strung with conspicuous orange buds. **Leaves** To 60 × 7 mm; dull green above with a frosting of the tiny silver scales that also *completely coat the underside*. **Flowers** Dioecious. Tiny clusters in spring; female trees (with longer, pointed buds) are wreathed from September to February in clusters of *orange berries* (too acid for most birds to eat) whenever a male is close enough to pollinate them.

COMPARE Oleaster (right); Silver Willow (p.88).

OTHER TREES *H. salicifolia* (the Himalayas) in a few big gardens: leaves to 70 × 12 mm, finely white-*felted* rather than scaly beneath. A sturdier tree, to 12 m, with *pallid yellow* berries.

Oleaster

Elaeagnus angustifolia

(Russian Olive) W Asia; naturalized in S Europe and long grown further N; very occasional in warmer parts.

APPEARANCE Shape Broad and spikily twiggy on a short leaning bole, to 11 m;

occasionally spiny. **Bark** Black-grey; shaggy with superimposed criss-cross ridges. **Shoots** Silver-scaled; buds inconspicuous. **Leaves** Larger than Sea Buckthorn's (p.227) – to 80 × 18 mm, but with the same silver scales. **Flowers** Hyacinth-scented bells (white outside, yellow inside), in early summer. **Fruit** 12 mm, yellow, with silver scales; sweet.
OTHER TREES *E. umbellata* (the Himalayas, China, Japan. 1830) can also reach tree size: leaves *rounded-oblong*, to 10 × 4 cm, *fresh green above* but with the characteristic silver scales beneath; its 1 cm fruits ripen from silver to red.

Dove Tree
Davidia involucrata

(Handkerchief Tree; Ghost Tree) One endangered species from W China. 1901. Occasional – larger gardens and some town parks in warmer areas; sometimes seeding. (Family: Cornaceae.)
APPEARANCE **Shape** Conic, broad-domed, or multi-stemmed, to 24 m. **Bark** Orange-brown; scaly flakes. **Shoots** Dark brown: like Black Mulberry's (p.140) but with *shiny crimson/green* buds. **Leaves** Elegant, heart-shaped; sunken veins and *big sharp teeth*; readily frost-blighted or hanging from drought-stress. *White-downy* beneath in var. *involucrata*; glaucous but *hairless* in var. *vilmoriniana* (*D. vilmoriniana*) – which is confined in the wild to high-altitude cloud-forests but is hardier and commoner in Europe, and otherwise alike. **Flowers** The 'doves', 'handkerchiefs' or 'ghosts' unfold yellow-green in late spring then hang in spectacular white tiers.

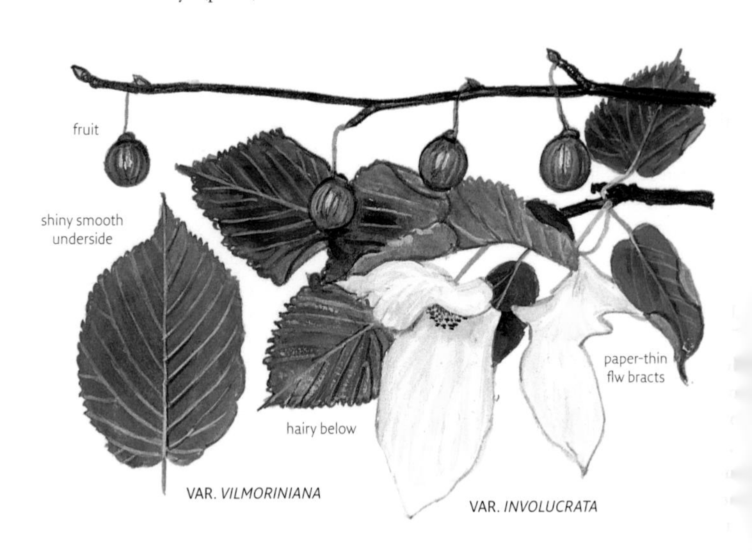

UCALYPTS

:alypts (500 Australasian species) include the world's tallest broadleaves and test-growing trees. Many mountain forms are still being tried out in N Europe, that trees outside the scope of this book can be found in unexpected places. The omatic oils can be irritant. The leaves are usually asymmetrical (curved) but the same each 'side'; typically untoothed and hanging vertically to minimize transpiration der a high sun. Like many Australian plants, the foliage alters as the plant matures: venile leaves' are relatively broad, seldom stalked and in opposite pairs; a few species ain this foliage and some lose it quickly, while many retain odd sprouts. Adult leaves narrower, stalked and appear alternate. All are evergreen; growth continues year-und. Flower-buds are usually carried in consistently sized clusters and are visible all ar long; they have 'lids' (a mass of fused petals) generally defined by a scar when an ter cap has been shed. The bud's 'lid' is pushed off by a fuzz of stamens. The fruit is a ody capsule containing fine seeds, dispersed through valves in the lid. Identification n be tricky: flower-bud clusters are the best feature to look for. (Family: Myrtaceae.)

INGS TO LOOK FOR: EUCALYPTS

Bark: How rough is it (at the base)?
Juvenile leaves: What shape are they? How glacous? Are they toothed?
Adult (alternate) leaves: Are they present? Are the main veins parallel with the midrib? Are they toothed? How long are they?
Flower buds: How many are there together? What shape are they? Are they stalked? In single or paired clusters? With distinct lids (what shape)? Green or glaucous?

EY SPECIES 

ue Gum (p.231): flowers usually borne gly. **Cider Gum** (right): short leaves; wers in 3s. **Broad-leaved Kindling Bark** 230): long leaves; flowers in 3s. **Shining ım** (p.231): long leaves; flowers in 4s–7s. **ow Gum** (p.232): main veins parallel th the midrib; flowers in 7s–12s.

Cider Gum
Eucalyptus gunnii

Marshy parts of the Tasmanian plateau and of the SE Australian Alps; its sap was once tapped for 'cider'. 1846. For long the most planted eucalypt in the UK (though scarcely the most ornamental) and abundant in warmer areas; often naturalizing freely.
APPEARANCE Shape Billowing: irregular, narrow branch-domes from an often long but *sinuous* trunk, or on several stems; rarely spire-like for long. To 35 m; potentially long-lived, but easily blown down at any age. **Bark** *Orange-grey*, or salmon-pink; sometimes white/grey. Soon rough with fine vertical flakes at the base; limbs smooth between long, hanging scrolls.
Shoots Pale green, then often vividly silver.
Juvenile leaves Round, silvery, 3–6 cm, finely *round-toothed*. **Adult leaves** (Soon very predominant.) *Oval*/willow-like, 4–15 cm, dark dull grey (rarely shining green); *seldom hanging vertically. Less scent*

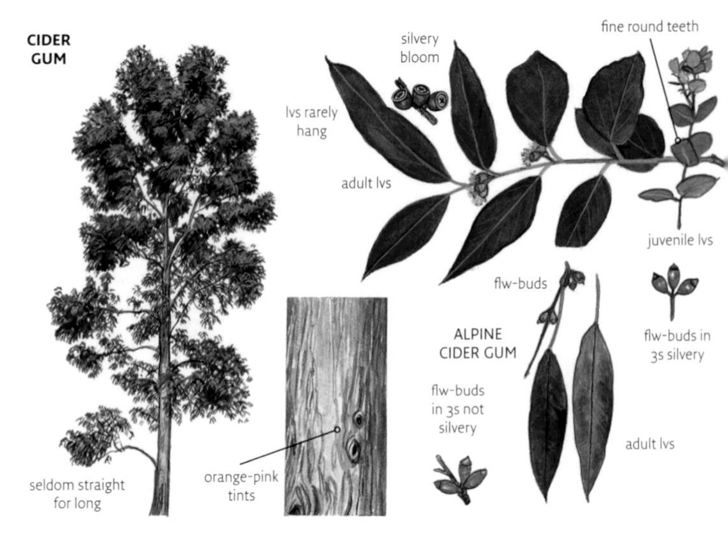

than most eucalypts'. Specimens with narrow leaves (12 × 3 cm) are sometimes 'Whittingehamensis', descendants of a famous old East Lothian tree. **Flower-buds** In 3s, variably *bloomed silver*; ice-cream-cone-shaped, the lids with sharp central knobs; fruit usually silver-bloomed.
OTHER TREES Alpine Cider Gum, *E. archeri*, a smaller high-altitude tree (rare to date) has usually greener, slender leaves; shoots, fruits *and flower-buds all apple-green, unbloomed.*

Broad-leaved Kindling Bark ✪ 💧
Eucalyptus dalrympleana

Tasmania, Victoria and New South Wales. Now locally frequent.
APPEARANCE Shape A towering dome, often on a straight bole; rarely a giant bush with sinuous stems. **Bark** Pale orange-brown (rarely cream); great shed lengths lodge in the branch-forks. **Juvenile leaves** *Pale green*, untoothed; rather *heart-shaped*, to 6 cm. **Adult leaves** Long (to 20 × 4 cm), *undulant* and hanging; flushi[ng] purplish, then dark matt grey. **Flower-bu[ds]** In 3s, green, their caps rather long-pointe[d]. **Fruit** To 9 mm.
COMPARE Shining Gum and Blue Gum (opposite).
OTHER TREES Ribbon Gum, *E. viminalis* (Manna Gum), is rarer, and has *boat-shap[ed]*

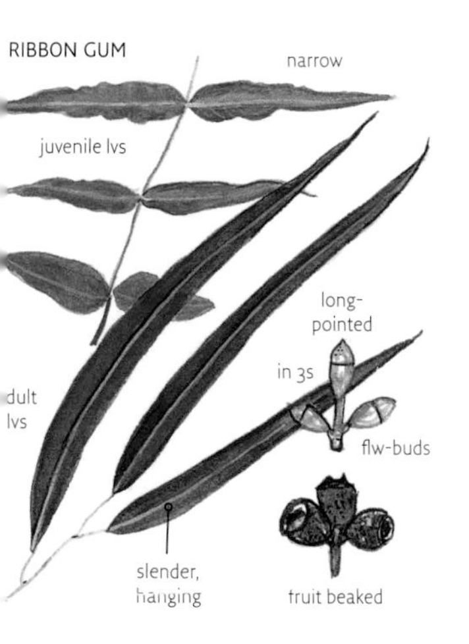

venile leaves (10 × 2 cm). Bark often
biter, with finer hanging strips, but rough
ar the base of old trees; leaves often
nderer, and a darker grey-green.

lue Gum  

calyptus globulus

smania (and rarely on the mainland).
29. One of the world's most planted
nber trees, but tender and confined in N
rope (as an old plant) to Ireland and the
e of Man, where it seeds freely: younger
es are very occasionally seen elsewhere.
PEARANCE Shape Conic, often on a
aight bole; then a towering, dense dome
44 m in Ireland). **Bark** Mostly smooth
ite and grey, with long spiralling strips.
Juvenile leaves *Large* (to 15 × 4 cm), close-set and *floppy*, and an intense silvery-white; predominant on trees until five years old and 10 m tall. **Adult leaves** Very long (to 30 × 5 cm), hanging and *dark* grey-green. **Flower-buds** Usually carried *singly: huge*, warty, and silvery. **Fruits** Large: *to 3 cm long*; carpeting the ground under old trees.
COMPARE Broad-leaved Kindling Bark (opposite); Shining Gum (below).

Shining Gum

Eucalyptus nitens

(Silver Top) Mountains of New South Wales and Victoria. Still rare, but one of the most magnificent of hardy eucalypts (and a fine timber tree); has grown 20 m in nine years in Argyll.
APPEARANCE Shape Rather *densely* conic so far; likely to reach 40 m. **Bark** Mostly smooth white and grey; hanging strips lodge

SHINING GUM

in the branch-forks. **Juvenile leaves** Big and floppy (like Blue Gum's; to 17 × 8 cm); rather glaucous. **Adult leaves** Hanging, long (to 25 × 4 cm), *very crescent-shaped*; fresh green to brilliantly blackish. **Flower-buds** In 4s–7s; small, *densely clustered*, with long conic caps. **Fruit** Small (6 mm), stalkless and shiny.

COMPARE Blue Gum (p.231) and Snow Gum (below).

Snow Gum ✪ ◐

Eucalyptus pauciflora ssp. *niphophila*

(*E. niphophila*) High mountains of Victoria and New South Wales. Occasional, but now more grown: perfect for small gardens.

APPEARANCE Shape A *dark, glistening dome* (cf. Shining Gum, p.231), on steep, thin limbs from a *short* bole; exceptionally to 25 m. **Bark** Sometimes white and satiny-smooth, or with vertical, *tight* grey strips; sometimes a multi-coloured jigsaw. **Shoots** Red; silver-bloomed. **Juvenile leaves** Rarely seen; oval, *not glaucous*. **Adult leaves** Carried by seedlings after four pairs; flushing red-brown, then glossy dark grey-green; 7–14 cm, oblong (broad or narrow); *plane and scarcely curved*, wit[h] *parallel main veins running right up them* and an abrupt, hooked tip. **Flowers** In *b[ig] clusters* (7s–12s); buds glaucous, *smoothly club-shaped* and upcurving.

Papauma
Griselinia littoralis

(Broad-leaf) New Zealand. *c.*1850. Frequent in mild parts, naturalizing: thrives on chalk and in salt spray. (Family: Cornaceae.)
APPEARANCE Shape Densely bushy (but to 20 m), on twisting, sprouty stems. **Bark** Dark brown; shaggy *curling scales.* **Leaves** Broad, *blunt*, to 11 × 8 cm; hairless, leathery. Usually matt and *pale apple-green*; some brightly variegated forms are now grown. **Flowers** Dioecious: little yellowy tails in spring. **Fruit** Blue-black berries, seldom seen in the UK.

Dogwoods (40 species) have flowers and fruit that are very diverse. Their leaves are untoothed, with elegantly incurved main veins, and usually opposite; if torn carefully in half they can remain held together by threads. (Family: Cornaceae.)

Things to Look for: Dogwoods

- Bark: What is it like?
- Leaves: Are they alternate? Evergreen? How many vein-pairs? How downy?
- Flowers, Fruit: What are they like?

Common Dogwood
Cornus sanguinea

(Cornel; *Thelycrania sanguinea*; *Swida sanguinea*) Europe, including England, Wales and Ireland. Abundant in downland scrub; frequent on heavy clays; now much planted. Skewers ('dogs') were made from the hard, straight twigs.
APPEARANCE Shape Broadly bushy, to 10 m; the leaves hanging elegantly, sometimes in tiers. **Bark** Grey; smooth, then with shallow, rounded ridges. **Shoots** Slim, straight and shiny (with some hairs at first): clear crimson in sun, lime-green in shade. **Buds** Scale-less: like *black bristles.* **Leaves** In opposite pairs; about 6 × 3 cm, with scattered stiff hairs on both sides; three or four vein-pairs. Rich *purple* in autumn.

Flowers In early summer: dull white, heavily scented heads. **Fruit** Purple-black 7 mm berries, their oil once used for soap and in oil-lamps.
COMPARE Purging Buckthorn (p.220): toothed leaves, scaly bark.

Strawberry Dogwood

Cornus kousa

(Japanese Cornel; *Benthamidia kousa*) Japan, Korea and (var. *chinensis*) central China. 1875. One of the most admired of all garden trees year-round, yet still very occasional.
APPEARANCE Shape Broad; rather tiered; to 10 m. **Bark** Thin *flakes of cream, orange and grey* (cf. Persian Ironwood, p.152). **Leaves** 4–8 cm, with three to four vein-pairs, elegantly hanging. Brilliant autumn colours. **Flowers** Through early summer, backed by four big creamy-white bracts. **Fruit** Edible: *magenta-red 'strawberries'* (cf. Bentham's Cornel).
VARIANTS 'Snowboy' has boldly white-margined leaves; 'Gold Star' has leaves centrally blotched in yellow (red towards autumn); both very rare to date.

Strawberry Tree

Arbutus unedo

Mediterranean region and SW Ireland, where it forms dense, low woods. Rather occasional in garden shrubberies in warmer parts; sometimes seeding. (Family: Ericaceae.)
APPEARANCE Shape Dense, low and rounded, to 15 m, on twisted limbs; a *more vivid* green than most evergreens.

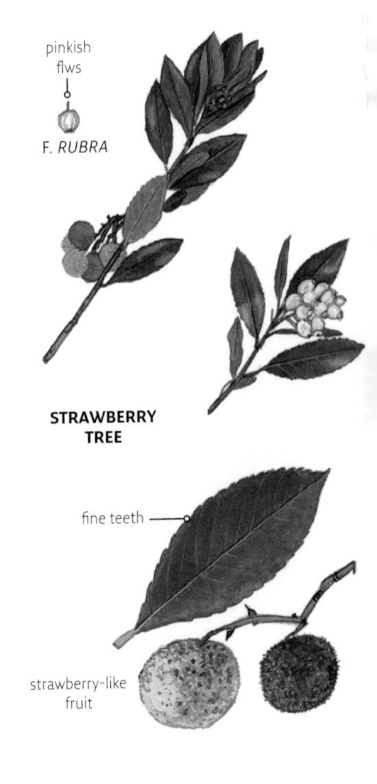

Bark Closely, shallowly scaly; *dull* red-grey. **Leaves** Small (8 × 3 cm), usually *serrated*, hairless but on downy stalks/young shoots. **Flowers, Fruit** Ivory-white bells in sprays *in autumn*, as last year's yellow 'strawberries' ripen scarlet. These tempt but disappoint ('unedo', the Roman name, means 'I eat one – only'), but make a good jam, and (in Portugal) the alcoholic drink *Medronho*.
COMPARE Phillyrea (p.241).
VARIANTS f. *rubra* has pink flowers; rare and rather ineffective.

Date-plum
Diospyros lotus

Cultivated across Asia and widely naturalized; long grown in the UK but now very occasional in old gardens. A tree in the Ebony family (Ebenaceae).
APPEARANCE Shape Often rather elegantly and narrowly domed, to 16 m, or on big, low stems; *very glossy, hanging leaves*. **Bark** *Blackish, with small oblongs* – sometimes like a bar of plain chocolate (cf. Common Pear, p.177). **Shoots** Green/brown, downy at first, with flattened, conic, 6 mm buds but no big end-bud. **Leaves** Deciduous and thin but looking rather evergreen, to 18 × 5 cm (though some are much smaller); untoothed; strange, heavy smell in dank weather (hints of lentil stew and wet dog). There are golden hairs *above the midrib*, scattered beneath, and on the *6–12 mm stalk*. **Flowers** Dioecious; 5–8 mm red/green urn-shaped bells in mid-summer (singly on female trees; up to three together on males). **Fruit** The 1–2 cm date-plums are purple or yellow, but in the UK never ripen sufficiently to become edible.

COMPARE Bay Willow (p.89). Black Cherry (p.194) and Almond (p.190): with serrated leaves. A puzzling plant, best told by its bark and very glossy, hanging foliage.
OTHER TREES Persimmon, *D. virginiana* (SE USA), is much rarer in the UK but very similar: leaves on *longer stalks* (10–25 mm) and flowers slightly longer (10–15 mm). The edible persimmons (4 cm; yellow flushed scarlet) again fail to ripen in the UK.

ASHES

Grouped in the same family as lilacs, jasmines and forsythias, ashes (60 species) have aerodynamic 'keys' like maples' (but symmetrical, and not paired) and opposite, usually compound leaves (in 3s on some strong shoots); they need a rich soil. (Family: Oleaceae

THINGS TO LOOK FOR: ASHES

- Bark: What is it like?
- Buds: Are they black when mature?
- Leaves: Is the leaf-stalk hairy? How many leaflets? (Are they hairy? Individually stalked?)

WENTWORTH ASH 'DIVERSIFOLIA PENDUL

Common Ash

Fraxinus excelsior

Europe (including Britain and Ireland); the Caucasus. Abundant/dominant except on light sands; planted everywhere. *Chalara fraxinea*, a fungus which has killed over 80% of Common Ash in parts of Europe, was first noted in Britain in 2012. The flexible 'ash-blonde' timber (used for tool-handles) burns even when green.

APPEARANCE Shape *Very open*; slender, cleanly curving limbs on an often long bole. The silvery shoots may droop then curl up like *branches of a chandelier*. Festooned frequently in ivy, which the airy crown fails to suppress, and affected by 'ash dieback' (probably due to a combination of environmental stresses): once known to 45 m; now rarely to 30 m. **Bark** Pale grey, developing a usually *regular* network of shallow, criss-cross ridges; rarely more like English Oak's. Erupting black bacterial cankers disfigure many trees. **Shoots** Grey.

COMMON ASH

late into leaf

details

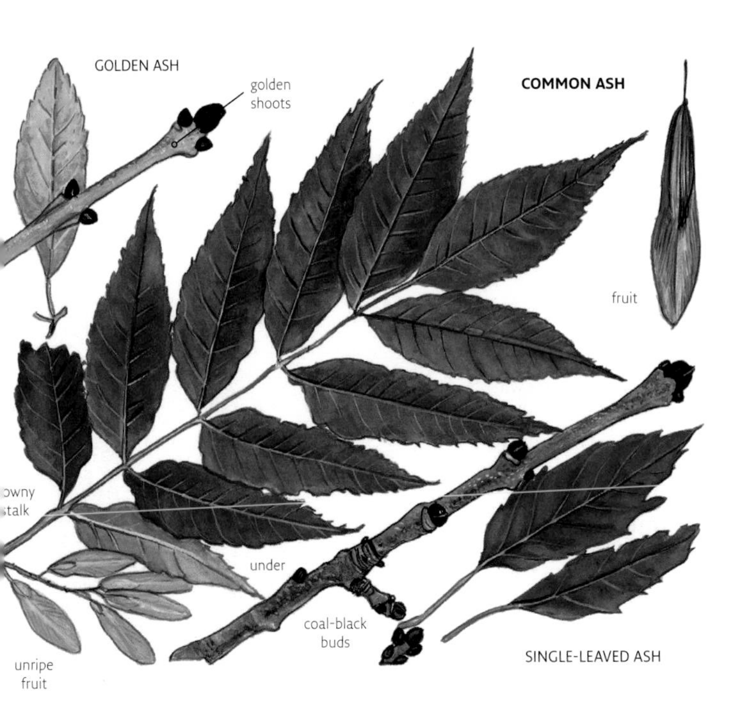

uds Mitre-shaped, soon *sooty-black* – other hes have brown buds. **Leaves** In opposite airs; *9–13* irregularly serrated leaflets (the de ones *stalkless*), dull above and *white-owny under the lower midrib*, on a slightly *owny main stalk*. The last wild tree into leaf, nd one of the first to go bare, fleetingly pale ellow. **Flowers** Nominally dioecious. Some ees change sex yearly, some carry branches f the wrong sex, some are hermaphrodite, nd some produce dual-sex ('perfect') flowers. **ruit** Bunched keys ripen biscuit-brown.

OMPARE Narrow-leaved Ash (p.238): ender, *hairless* leaflets. Claret Ash (p.238): ender leaflets on *hairless main stalks*. Manna Ash (p.239): smooth bark; 5–9 eaflets. Red Ash (p.240): 7–9 glossy leaflets. Elder (p.245): 5–7 leaflets.

ARIANTS Weeping Ash, 'Pendula', has traight shoots, like a shower of sleet, from ngular branches; frequent, to 17 m.

Wentworth Ash, 'Pendula Wentworthii', is rare but taller: the long-hanging shoots twist then *curl up at their tips*.

Golden Ash, 'Jaspidea', has *gold* not silver shoots for up to five years and brighter yellow autumn colours; a *densely leafy*, very slightly yellowish dome in summer. Occasional and vigorous; old trees (with little shoot-growth) often only noticeable from being grafted. 'Aurea' is much rarer: *stunted, twiggy growth*; smaller (green) leaves in summer.

Weeping Golden Ash, 'Aurea Pendula', is rare: a slow-growing, flat, gaunt plant with stiffly and shortly hanging golden shoots.

Single-leaved Ash, f. *diversifolia*, is occasional and has *one big undivided leaf* (rarely three leaflets; cf. Box Elder, p.214), *jaggedly toothed/lobed* and to 20 cm. A puzzling graft (rarely found wild), but with typical bark *and black buds*. The weeping 'Diversifolia Pendula' is very rare.

Narrow-leaved Ash

Fraxinus angustifolia

W Mediterranean, N Africa. 1800. Occasional.

APPEARANCE Shape Denser than Common Ash (p.236), to 30 m; upswept or slightly weeping; ethereal, *feathery clouds* of shiny foliage. Old trees are often grafts on a stock of Common Ash – which they outgrow bizarrely. **Bark** Dark grey, soon with quite *deep, rugged ridges*. **Buds** Brown, finely grey-woolly. **Leaves** With 7–13 slender (60 × 15 mm) leaflets, shiny, stalkless and *completely hairless*, on *hairless* main stalks. **Flowers** As Common Ash.
COMPARE Red Ash (p.240); Claret Ash (below).

Claret Ash

Fraxinus angustifolia ssp. *oxycarpa* 'Raywood'

An Australian clone of the Caucasian subspecies of Narrow-leaved Ash. By 1925. Now abundant as a park and street tree: breathtaking in autumn as the shoot-tips turn *royal purple* and the interior follows with orange, pink and gold.

APPEARANCE Shape Openly globular, to 25 m, with narrow forks which are the tree's main fault – the limbs shear off easily. **Bark** Dark grey; *smooth for many years* then finely ridged. **Leaves** With usually seven *slender*, feathery, glossy, rich-green, stalkless leaflets with a *band of hairs* under the base of the midrib, on *hairless* common stalks.

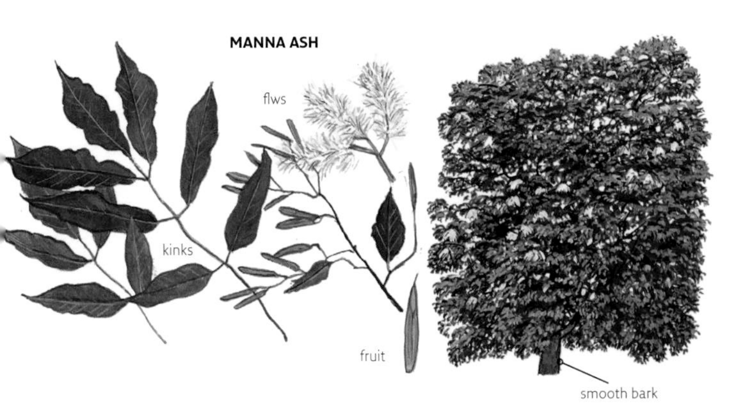

1anna Ash

raxinus ornus

Europe; Asia Minor. Long grown in the JK: very frequent in parks and streets. **PPEARANCE Shape** A *dense, leafy dome* n rather twisting branches; rarely to 7 m. **Bark** Dark grey, usually *very nooth*; it has been tapped for a sugary nanna'. **Buds** Grey-brown, woolly. **eaves** With five to nine very dark, broad, ather glossy leaflets, hairy under the lower nidrib and on their *individual 1 cm stalks*; ommon stalk *kinky*, with tufts of hair at ach junction. Muted yellow and *deep pink* utumn colours. (Var. *rotundifolia*, from S taly and the Balkans, has leaflets only 3 cm ong.) **Flowers** Insect- not wind-pollinated: *'ense, fluffy plumes* of very narrow 1 cm reamy petals, *showy* in late spring.

Vhite Ash

'raxinus americana

: North America. 1724. Occasional in varmer areas.

PPEARANCE Shape Often gaunt: steep ranches, straight shoots. To 30 m. **Bark** 'ale grey; *shallow* ridges. **Shoots** Hairless; eaf-scars have *concave* upper edges. **Buds** Brown-woolly, the end ones *blunt*. **Leaves** Vith seven or nine *large* (to 15 × 7 cm)

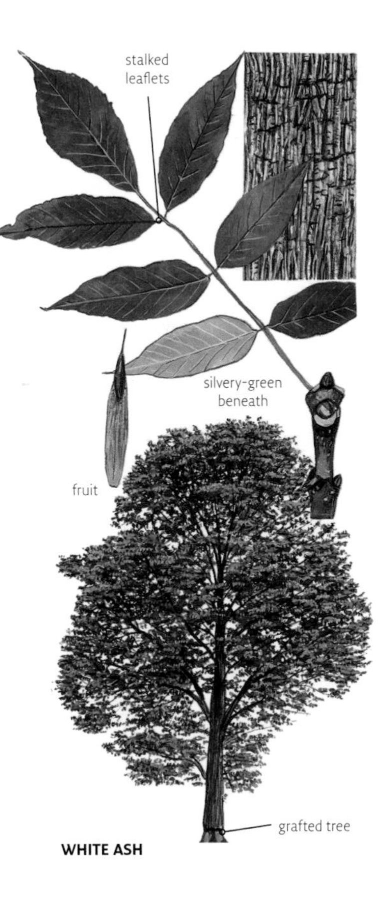

leaflets with distant teeth and *5–10 mm individual stalks*; often hairless but *whitish beneath with papillae* (waxy warts about the size of the ink-dots on this page); main stalk *hairless*. Butter-yellow in autumn. **Flowers, Fruit** As Common Ash (p.236).
COMPARE Red Ash (below): differences emphasized.

Red Ash

Fraxinus pennsylvanica

(Green Ash) E North America. 1783. Locally occasional (town parks, streets).
APPEARANCE Shape Often domed (to 23 m); like a *glossy*, well-groomed Common Ash (straighter shoots; brighter autumn golds). **Bark** Usually grey-*brown* with *close, sharp ridges* (like Norway Maple's, p.208), which may become shaggy, like some hickories'. **Shoots** Sometimes palely velvety at first; the leaf-scars have *level* upper edges. **Buds** Brown-woolly, the end ones *pointed*. **Leaves** With three to nine leaflets, large (to 14 × 6 cm) or slender and long-pointed with few jagged teeth or none (or in odd trees forking/jaggedly lobed); the lower leaflets, at least, have *5–8 mm individual stalks*; the grey-green under-leaf *lacks papillae* but has hairs (reduced, in the glossy-green widespread form distinguished as 'Green Ash', and once as var. *lanceolata*/*F subintegerrima*, to tufts in the lower vein-joints). The main stalk may be downy.
COMPARE White Ash (p.239): differences emphasized.
VARIANTS 'Variegata' is richly white-variegated; very rare and apt to revert.

Chinese Tree Privet

Ligustrum lucidum

Central China; long cultivated as the food-plant of a wax-producing aphid. 1794. Locally occasional.
APPEARANCE Shape Neatly domed, often on many closely forking stems; to 18 m; crown sparser than most evergreens'. **Bark** Grey-buff; fluted and sprouty; distant fissures. **Leaves** In opposite pairs; to 15 cm; thick, long-pointed, hairless and *brilliantly glossy* (matt beneath). **Flowers** Yellow-white spires decking the tree from mid-summer into winter, when next year's green heads

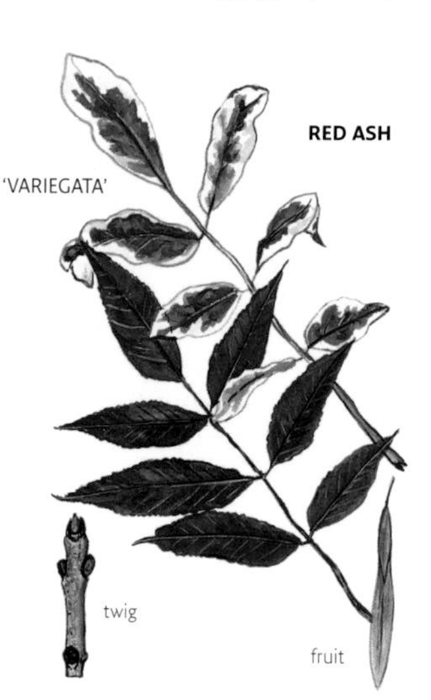

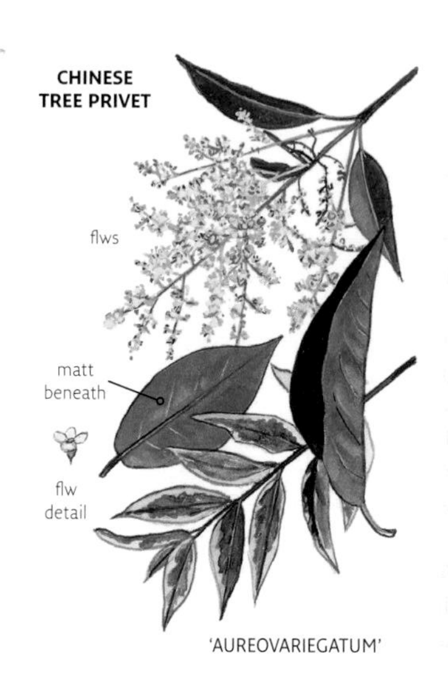

evelop. **Fruit** Blue-black, 1 cm berries
oisonous, but with medicinal uses); rare
the UK.

OMPARE Bay (p.150): duller, alternate
aves.

ARIANTS 'Excelsum Superbum' is as
agnificent as its name: bright double
ariegation of *gold and cream* so the crown
lows pale yellow. Very occasional. 'Tricolor'
rare) has narrow silver leaf-margins, *pink* at
rst. 'Aureovariegatum' (very rare) has a *dull,
reeny-gold* variegation.

hillyrea

hillyrea latifolia

Incorporating *P. media*) Around the
Mediterranean; long grown in warmer parts
f the UK but now very occasional.

PPEARANCE Shape A low, dense, *blackish*
vergreen dome; to 11 m. **Bark** Grey-black;
losely but roughly square-cracked. **Shoots**
With tiny hairs at first. **Leaves** In opposite
airs; hairless and glossy above; typically
× 2 cm, with neat little rounded teeth.
lowers Dioecious: small green-white
lusters early in summer. **Fruit** Females
arry blue-black, 1 cm berries.

COMPARE Holm Oak (p.123): Phillyreas, when noticed at all, are regularly passed over for this. Strawberry Tree (p.234).

VARIANTS f. *spinosa* has strongly toothed leaves (cf. Holly 'Myrtifolia'); very rare. f. *buxifolia* has small, untoothed leaves – very like Box's (p.200), but with the type's dark bark; also very rare.

Things to Look for: Foxglove & Bean Trees

- Bark: How fissured is it?
- Leaves: What shape are they (especially at the tip and the base)? How hairy beneath? Is the leaf-stalk downy? What colour is it?

Foxglove Tree

Paulownia tomentosa

(Empress Tree; *P. imperialis*) N China – where only an empress could have one on her grave. 1838. Occasional in warmer areas. Glorious in flower (weather permitting), but sometimes cut annually as a foliage plant, regrowing 3 m shoots – hollow, like postage tubes – and elephantine leaves. One of a single small genus of trees among the many herbs of the Foxglove/ Figwort family (Scrophulariaceae).

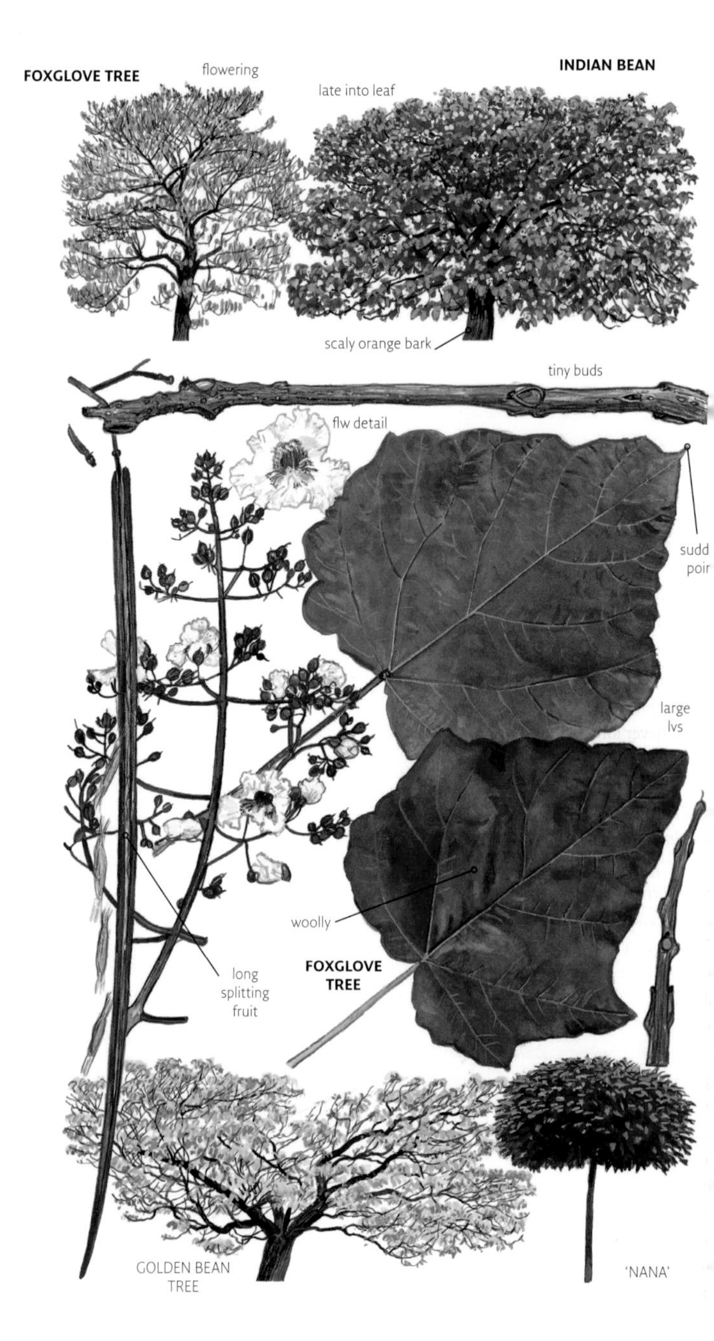
FOXGLOVE TREE
flowering
INDIAN BEAN
late into leaf
scaly orange bark
tiny buds
flw detail
sudd
poir
large
lvs
woolly
long
splitting
fruit
FOXGLOVE
TREE
GOLDEN BEAN
TREE
'NANA'

PPEARANCE **Shape** Wide-spreading then pcurving branches from a straight trunk; 26 m, but rarely even 15 m. Fragile, open nd short-lived; may sucker. **Bark** *Grey*, nely roughened; very *shallow, wide, rounded* dges in age. **Shoots** Stout, pink-brown, nely scurfy. Winter shoots end with three ny purplish buds round a dead tip (which as failed to ripen) or – older plants – in ndelabra of huge brown-woolly flower-uds. **Leaves** Gigantic (to 35 cm) and flimsy, eart-shaped and sometimes lobed/coarsely othed; *softly woolly on both sides*, on a very *oolly stalk*. **Flowers** In late spring, but efore the leaves: tall heads of deep mauve, ighly fragrant bells. The buds – developing e summer before – are vulnerable both to inter cold and late frosts.

OMPARE Hybrid Bean Tree (p.244).

THER TREES *P. fargesii* (W China, *c.*1896), very rare: *almost hairless leaves* (cf. bean ees, with rugged barks) and slightly paler owers.

P. fortunei (S China, Taiwan, E Himalayas; 1940) is also very rare: slenderer, darker aves, never lobed; flowers lilac outside and *eam inside* with purple splashes.

ndian Bean

atalpa bignonioides

Southern Catalpa) Georgia, Florida, labama and Mississippi. 1726. Quite frequent in hot areas. (Family: ignoniaceae.)

PPEARANCE **Shape** Typically very broad nd open: level, twisting branches on a short, anted bole; to 18 m. **Bark** *Shallowly scaly/ quare-cracked; orange to pink-brown* (rarely reyish). **Shoots** Stout, grey-brown, soon airless, with 1.5 mm orange buds – often in approximate) 3s, the end ones surrounding he scar of a dead shoot-tip (cf. Foxglove Tree), and helping to make the tree look ead for eight months of the year. **Leaves** Big (15–30 cm) and foul-smelling when ruised; rather *pale fresh green; rounded* or very shallowly heart-shaped at the base and *tapering quickly to a sudden 1 cm point*; on saplings frequently but on mature trees *rarely* with small side-lobes. Closely hairy beneath; stalk green and *hairless*. They emerge briefly purplish and fall without autumn colour. **Flowers** In big candles at the growth-tips in late summer: white with yellow and purple splashes. **Fruit** 'Bean pods', to 40 cm, dangle after hot summers.

COMPARE Other catalpas (pp.244–5).

VARIANTS Golden Bean Tree, 'Aurea' (locally occasional) has *rich yellow* foliage, fading to soft green by autumn; usually conspicuously grafted. Shy-flowering. 'Nana' is rare and is grafted on to a stem of the type, producing a frizz of short, weak stems, with smaller leaves. (Compare, out of leaf, Mop-head Robinia, p.199.)

Western Catalpa

Catalpa speciosa

(Northern Catalpa) Central USA: Arkansas to SW Indiana. 1880. Occasional in warm areas; less vulnerable to frost damage than Indian Bean but still needs plenty of summer heat to ripen its wood.

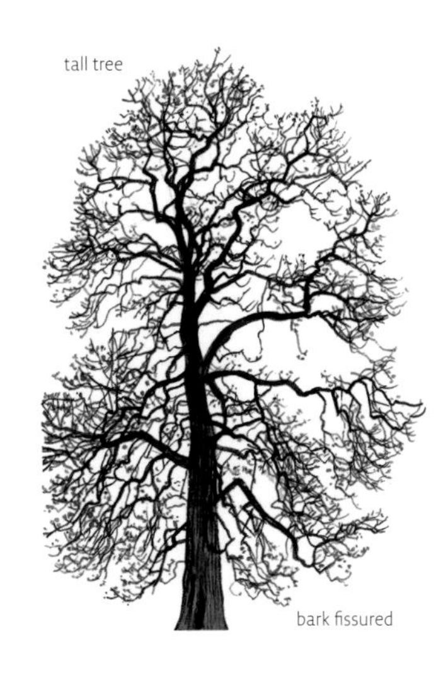

APPEARANCE Shape A *tall, densely leafy* dome of short, twisting branches, to a narrow top (20 m); bole often straight. **Bark** *Grey*; quite *deep*, scaly vertical ridges. **Leaves** Longer and darker than Indian Bean's, and scentless; usually *deeply heart-shaped at the base* and tapering *narrowly to their long apex; rarely* on mature trees with small side-lobes. Closely brown-hairy beneath; stalk finely hairy at first. **Flowers** A little ahead of Indian Bean's, in large but *open* heads. **COMPARE** Indian Bean; Hybrid Bean Tree (below).

Hybrid Bean Tree

Catalpa × erubescens 'J.C. Teas'

(Indian Bean × Yellow Catalpa; raised by J. C. Teas in Indiana in 1874; introduced 1891.) Occasional in warm areas.

APPEARANCE Shape Wide, tall (to 20 m) and rather *open*; trunk often sinuous. **Bark** Grey-brown, with quite deep scaly ridges (like Western Catalpa's (above) – but less predominantly vertical). **Leaves** Very big (to 35 cm), deep green; heart-shaped at the base, *often as broad as long; 20–50 per cent* (on older trees) with small side-lobes. **Flowers** A little later than Indian Bean's; in slightly less dense candles, but richly fragrant. **Fruit** 40 cm 'bean-pods' (filled only with fluff) hang on profusely in warm areas. **COMPARE** Western Catalpa (above); Yellow Catalpa (below).

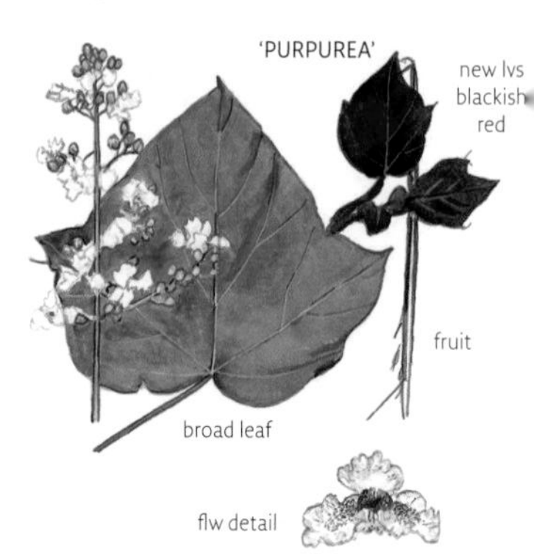

VARIANTS 'Purpurea' has strongly purple-splashed flowers (creating a *lilac* tinge), and smaller, duller, darker leaves which, for several days as they open, are *more intensely black-purple* than other bean trees', and they retain some purplish veins and *purple stalks*. Rare.

Yellow Catalpa

Catalpa ovata

China (and Japan?). 1849. Very occasional in warm areas.

APPEARANCE Shape Rounded, or slender on a *long* sinuous bole. To 22 m. **Bark** Grey-brown; scaly ridges. **Leaves** *Matt, dark yellowish green*, smaller than Indian Bean's (but to 25 cm); *usually with big lobes* and at least as broad as long; on *dark red* stalks with short, stiff hairs. Purple glands under the leaf-base ooze nectar. **Flowers** Appearing with Indian Bean's; white, with yellow staining and red spots – appearing dull *creamy yellow* at a distance. In small,

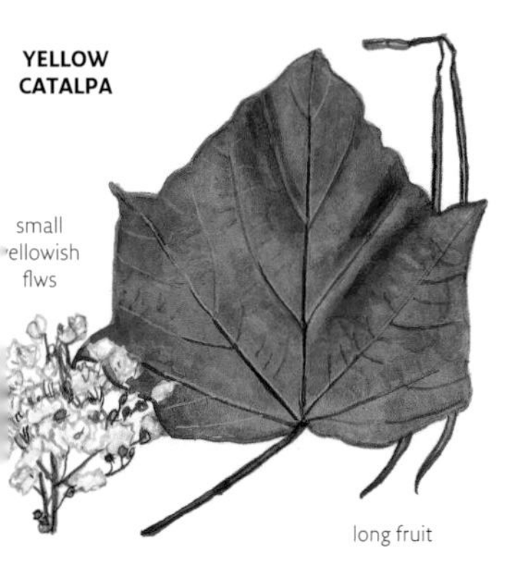

arse heads but deliciously scented
trawberries, with a hint of soap flakes –
ke Silver Lime, p.225).

lder

ambucus nigra

urope (including Britain and Ireland); Africa; SW Asia. Ubiquitous except n sands; scarcely planted. (Family: aprifoliaceae.)

PPEARANCE **Shape** Branches arch over ith vigorous erect sprouts; to 10 m. **Bark** *reamy-grey*: corky criss-cross ridges. **Shoots** ream-grey, with *raised warts and often andering tips*. **Buds** Purplish; *spiky-scaled, like pineapples* – expanding in mid-winter, soon after the last lemon-yellow or pink leaves drop. **Leaves** In opposite pairs; *five or seven* (rarely three or nine) leaflets, dull, with some stiff hairs. **Flowers** In *creamy plates* in early summer, heavily scented; sometimes made into 'champagne'. **Fruit** Elderberry wine comes from the black berries – 'the Englishman's grape' – which ripen in early autumn and are poisonous raw (like the leaves), but have many medicinal uses.

VARIANTS Parsley-leaved Elder, f. *laciniata*, is very feathery and rare; less cut-leaved sports are occasional. Golden Elder, 'Aurea', is much grown. (The gold cut-leaved 'Plumosa Aurea' belongs to a Continental scarlet-berried bush, *S. racemosa*.)

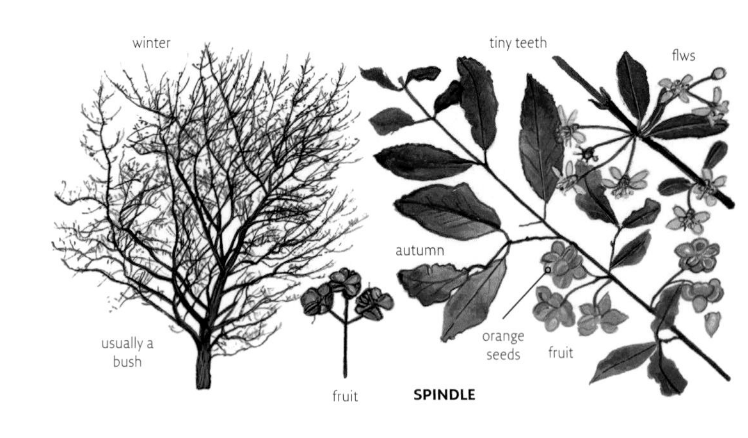

SPINDLE

Spindle

Euonymus europaeus

Europe (including Britain and Ireland) to the Caucasus. Abundant on richer soils. The straight, hard twigs were used for skewers and spindles. (Family: Celastraceae.)

APPEARANCE Shape Generally bushy; to 9 m. **Bark** Deep green, with faint fawn 'snakes'; then grey-brown, closely grooved. **Shoots** Thin, straight, slightly squared and *deep green*. **Buds** Greenish, short-conic. **Leaves** Blackthorn-shaped (but *opposite*), to 8 cm, with *minute sharp teeth*; a little shiny and waxy, hairless. **Flowers** In small creamy heads in early summer. **Fruit** Poisonous *orange berries, from four magenta lobes*, can be delicately spectacular among leaves of th same colours.

COMPARE Purging Buckthorn (p.220).

OTHER TREES *E. bungeanus* (NE Asia, 1883; very rare) is perhaps the handsomest *tree-sized* spindle (domed, to 10 m): broader hairless leaves (to 10 × 6 cm) hang gracefully. Fruit-lobes pale creamy-pink.

MONOCOTYLEDONS

Cabbage Palm

Cordyline australis

(Cabbage Tree; *Dracaena australis*) New Zealand. 1823. Abundant in milder, coastal areas; seeding freely and suckering. (Family: Agavaceae.)

APPEARANCE Shape Branching after flowering (typically at 5 m and about eight years); ultimately densely funnel-shaped, to 16 m: one of the few monocotyledons (plants, like grasses and lilies, which on germinating grow one 'seed-leaf' not two, and have parallel, not networked, leaf-veins) that steadily builds a thick, branching bole. **Bark** Creamy-grey; rather corkily square-cracked; whiskery at first from shed f-bases. **Leaves** Linear, to 90 × 8 cm; very ugh and can be made into paper, though ung sprouts are eaten in New Zealand; ing ones may phosphoresce in the wild. : Yuccas – bushier plants, with towers creamy-white big bell-flowers – which have shorter, thicker, greyer leaves.) **Flowers** Fragrant, in huge airy white plumes through early summer. **Fruit** Blue-white 6 mm berries. **VARIANTS** Many purple, brownish and prettily variegated clones are now in commerce.

the palm family (Arecaceae), there are about 3000 species. Their leaves and owers usually form from a single bud at the top of a stem that increases in height it not in diameter.

anary Palm

oenix canariensis

demic to the Canary Islands but widely ltivated; scarcely hardy in the UK, but w frequent as a young plant in warm eas.

PEARANCE Shape An amazing star-burst arching and hanging blue-green leaves, a bole sturdier than Chusan Palm's elow). **Bark** Tiled with clean, bright own leaf-stumps. **Leaves** To 7 m n saplings necessarily much aller), the *parallel leaflets* lded upwards at their bases. **THER TREES** Date Palm, *P. ctylifera*, has similar leaves ss drooping). It is easy to ise from date-seeds d even grows up in bbish heaps, but ils to survive N ropean winters.

P. theophrastii (with suckering habit, like Date lm's) is endemic to Crete; not rdy in the UK.

Chilean Wine Palm, *Jubaea ilensis*, an ally of Coconut Palm, as represented in the nineteenth ntury at Kew Gardens by a large ee near the Main Gates, but is scarcely own now and considered tender; rare in hile, as tapping the sugary sap involved felling the tree. The trunk, with *smoothly wrinkled, pale grey bark*, can be 1 m thick; leaves very like Canary Palm's but on thicker, greyer central stems.

Chusan Palm
Trachycarpus fortunei

(Windmill Palm) China; N Vietnam; Upper Burma. 1836. The hardiest palm, occasional in milder areas, where it seeds freely.
APPEARANCE Shape To 15 m. **Bark** Covered in the matted fibres of old leaf-bases, like an ancient hot-water pipe (and for much the same reason: the lagging helps keep frosts out); highly inflammable and something of a liability in public parks. Rarely smooth and green near the base. **Leaves** *Fan-like*, the tips usually shredded and hanging. The 1 m stalk has tiny spines **Flowers** In huge (to 60 cm) egg-yolk-yello drooping plumes. **Fruit** Blue-black, 1 cm berries; rare in the UK.
OTHER TREES Dwarf Fan Palm, *Chamaero humilis* (W Mediterranean, 1731), is rare: 5 m, suckering and usually on *several stems* Leaves smaller, stiffer, greyer, on *viciously spined* stalks; flower-heads (dioecious) only 15 cm high.

GLOSSARY

ternate: Not opposite each other.

ther: Pollen-bearing male part of flower, tipping a stamen.

pressed: Lying (almost) flat against.

ricle: Backward-pointing lobe at base of leaf (or petal).

ckcross: Offspring of a hybrid crossed with one of its parents.

oom: Waxy surface layer, easily rubbing off.

act: Modified leaf, behind a flower/flower-head/seed.

d: Embryo leaf, flower and/or shoot, plus any protective covering.

nker: Fungal/bacterial infection, causing bark lesions.

tkin: Tail-like single-sex flower-head, with scale-like bracts and stalkless flowers.

himaera: Plant originating in the non-sexual fusion of two species' tissues.

hlorosis: Sickness due to a failure to obtain iron from alkaline soils, leading to yellow foliage.

lone: All the plants propagated vegetatively from an individual.

olumnar: Narrow, erect, with parallel sides.

ompound: Made up of several parts (e.g. leaflets).

onic: Narrowly pointed from a broad base.

oppice: To fell a tree so that the base sprouts to make new trees; a stump treated in this way; to regrow like this.

rown: A tree's whole structure above the trunk.

ultivar: A '**culti**vated **var**iety' or mutant clone, nursery-distributed.

ED: Dutch Elm Disease: fungal pathogen of plants in the elm family (researched largely in Holland).

Dioecious: With male and female flowers on separate plants.

Distant: Widely separated.

Double (flowers): With some or all stamens mutated into extra petals.

Doubly compound: Compound, with each leaflet itself compound.

Doubly toothed: With serrations that are themselves toothed.

Down: Covering of tiny, soft hairs.

Endemic: Native to one small area.

Family: Scientific grouping of plants, divided into genera and into which orders are divided.

Fastigiate: Growing nearly vertical branches and/or shoots.

Flush: To come into leaf.

Forma (plural formae): Scientifically described sport or minor variant of a species.

Free: Not attached; projecting.

Genus (plural genera): Scientific grouping of species, into which families are divided.

Gland: Secreting organ: a tiny, hard, often sticky swelling.

Glaucous: Pale bluish grey (usually due to a bloom).

Graft: To fuse a shoot of one kind of tree on to roots of another; a tree fused in this way.

Habit: Manner of growth.

Hybrid: The offspring of different species, subspecies or varieties.

Impressed: Sunk below the surface.

Layer: A branch touching the ground and rooting; the secondary trunk that results.

Leader: Top shoot of a tree adding height.

Leaflet: Discrete part of a compound leaf.

Lenticel: Raised breathing-pore on a shoot, trunk or fruit.

Lobe: Promontory on a leaf (or petal, etc.), larger than a tooth.

Microspecies: A species which is usually local and produces fertile seed without pollination.

Midrib: A leaf's central main vein.

Needle: Linear leaf of a conifer.

Opposite: In a pair, one on either side of a shoot.

Original: The first tree of a taxon in cultivation in a country.

Petal: Segment of a flower's inner whorl.

Plane: Flat-surfaced.

Pollard: To cut a tree at 2–4 m so that it grows new limbs; a tree cut in this way.

Scale: A flattened appendage, not leaf-like or petaloid.

Scion: Shoot grafted on to a stock to make a new tree.

Semi-double (flower): With a few but not many stamens mutated into extra petals.

Sepal: Segment of a flower's outer whorl (outside/behind the petals and persisting, withered, on top of some fruits).

Serrated: With many teeth (which point forwards).

Shoot: Young twig.

Simple: Not double or not compound.

Sinus: Recess between two lobes.

Sport: A genetic freak.

Spur: Short side-twig, growing very slowly each year.

Stamen: Male organ in a flower: an anther and its stalk.

Stipule: Appendage, usually leafy, at the base of a leaf- or flower-stalk.

Stock: Roots of a plant on to which a scio is grafted.

Style: Female organ in a flower: a stalk for the stigma which receives the pollen.

Subspecies: A scientifically recognized (regional) variant of a species.

Sucker: A sprout from a root, which may grow into a new tree.

Taxon (plural Taxa): A single discrete nomenclatural unit.

Tepal: Undifferentiated sepal/petal in som flowers (e.g. Magnolias, Tulips).

Tooth: Promontory on the margin of a lea or petal, smaller than a lobe.

Trifoliate: With leaves divided into three leaflets.

Type: The usual form of a species (as opposed to a variety or clone); the individual on which a taxon's scientific description was based.

Variegated: With leaves consistently lacking green coloration in patches.

Variety: A scientifically recognized (morphological) variant of a species.

Whorl: Circle of branches, leaves, flowers or other organs.

Wing: Flat membranous extension.

Woolly: Covered in long, soft hairs

NDEX

T

U

W

X

Y

Z